Foundation Discrete Mathematics
for Computing

Tutorial Guides in Computing and Information Systems

Series Editors

Professor David Howe, De Montfort University
Dr Martin Campbell-Kelly, University of Warwick

About the series

The Tutorial Guides in Computing and Information Systems series covers the first and second year undergraduate programme and the Higher National Diploma courses in these subjects. The essentially practical nature of the books is particularly appropriate to today's courses where examinations reflect an increasing emphasis on problem-solving skills. The books are characterized by a high proportion of worked examples, practical exercises and frequent friendly tutorial-style notes in the margins of the text. The problem-solving approach also addresses the requirements of practitioners in business and industry.

Each book stands alone in its subject area; the series as a whole provides comprehensive coverage of the first two years of the undergraduate or Higher National programme.

About the Series Editors

David Howe is Head of Department of Information Systems at De Montfort University. He specializes in the analysis and design of information systems and is the author of *Data Analysis for Database Design*.

Martin Campbell-Kelly is Senior Lecturer in Computer Science at Warwick University. He teaches a number of subjects of the undergraduate programme and is a recognized authority on the history and development of computing.

Other titles in the series

Z: A Beginner's Guide
David Rann, John Turner and Jenny Whitworth

Introduction to C++
David Dench and Brian Prior

Human Computer Interaction for Software Designers
Linda Macaulay

The Relational Database
John Carter

Programming in C
John Gray and Brian Wendl

Information Systems: Strategy to Design
Chris Clare and Gordon Stuteley

Systems Analysis Techniques
Barbara Robinson and Mary Prior

Foundation Discrete Mathematics for Computing

Dexter J. Booth
School of Computing and Mathematics
The University of Huddersfield
UK

Springer-Science+Business Media, B.V.

First edition 1995

© 1995 Dexter J. Booth
Ursprünglich erschienen bei Chapman & Hall 1995
Typeset in Great Britain by Columns Design & Production Services Ltd, Reading

ISBN 978-0-412-56280-8   ISBN 978-1-4899-7114-2 (eBook)
DOI 10.1007/978-1-4899-7114-2

A catalogue record for this book is available from the British Library

Library of Congress Catalog Card Number: 94–68783

♾ Printed on permanent acid-free text paper, manufactured in accordance with ANSI/NISO Z39.48-1992 and ANSI/NISO Z39.48-1984 (Permanence of Paper).

Contents

Preface

The computer is a perfect example of the physical realization of a mathematical idea, and to properly understand the workings and rationale of a computer requires a familiarity with mathematics generally and discrete mathematics in particular.

Mathematics as its own discipline is more concerned with form than content; the mathematician is more interested in the pattern than with the elements that create the pattern. However, we are not all mathematicians, and to us the elements are every bit as important as the pattern and, in some cases, more so. Content is essential for our proper use of a computer, but form is essential for our proper understanding of a computer.

This book treats both form and content hand in hand, the form being continually displayed and reinforced by the structure of the book and the content exemplified and reinforced by the worked examples and exercises that are used throughout to demonstrate the objectives.

The purpose of this book is to expose the novice to the foundations of discrete mathematics as applicable to computing. Such mathematics is concerned with a formalized way of looking at the working of a computer, and in its study we are concerned with two distinct aspects. Firstly one must consider the mathematics as a set of abstract concepts and, secondly, one must consider the application of mathematics to describe the physical working of the computer. Most texts attempt to look at both of these aspects side by side on an equal footing, but in this text the material is intended to concentrate more on the former with the intention of enabling the reader to develop the frame of mind that permits application to be viewed in an informed light. Throughout the book illustrative examples are given followed by exercises in the identical format. By following the working of the examples and then attempting the exercises it is hoped that the reader will develop an appreciation of the mathematics and become prepared for an exposure to software engineering and other mathematics books that deal with items in this book in greater depth and detail. A bibliography of recommended texts for further reading is given at the end of the book.

Discrete mathematics, apart from set theory, is not the kind of mathematics that is taught in high school. It is quite different and, indeed, you may be forgiven for having the feeling that it is not really mathematics at all. If this attitude helps you then fine – just look for the patterns and enjoy.

Dexter J. Booth
The University of Huddersfield

Acknowledgements

I wish to dedicate this book to Olivia and James and to extend my grateful acknowledgements to Dr Adrian R. Jackson and Dr John K. Turner for their early contributions during the development of this book.

Part One

Foundations of Logic

When you sit in front of your TV and press the 'ON' button of your remote control the TV 'comes to life' and a picture appears. You don't like the programme so you press a channel button and the channel changes. All familiar stuff, but halt awhile and think about what exactly it is that you are doing.

You are communicating with your TV and we all know about communication – we do it all the time. We ask the milkman to deliver two pints of milk each day and he does. We telephone the plumber and she comes to mend the leak behind the washing machine. We press the 'ON' button on the TV remote control device and the TV comes on.

In each case we have communicated a wish and that wish has been acted upon. Let us not use the word wish; instead we shall use the word **instruction**.

In all three cases mentioned – the milkman, the plumber and the TV – we issued an instruction and the instruction was obeyed. The means whereby the instruction was communicated is not important for now; what is important is that we understood the instruction given and the recipient understood the instruction received.

Human understanding is something that grows with us. We learn by experience and by interacting with other people. Machine understanding is rather different. A machine can understand only what it was designed to understand – your TV was designed to understand what is meant when various buttons are pressed on the remote control, but you try shouting at it. No matter how loud you shout you will elicit no response – it was not designed to be shouted at.

In order to design a machine that understands instructions we must first have a method of describing the meaning of instructions, and this is what Part One is all about – the rules that govern our understanding of meaning.

The rules are not new. They were understood by the Ancient Greeks, who used them to study the rules of rhetoric – the art of a good argument. However, it was not until George Boole codified the rules into a coherent system in the nineteenth century that they became generally available for use in a form other than for conversation.

Chapter 1

Propositions: either true or false

OBJECTIVES

When you have completed this chapter you will be able to:

- □ distinguish propositions from other types of sentence;

- □ construct compound propositions using the logical connectives AND, OR and NOT;

- □ construct truth tables for compound propositions;

- □ demonstrate the logical equivalence of compound propositions by equating truth tables;

- □ translate propositions into symbolic form;

- □ construct expressions for switching circuits;

- □ construct truth tables for switching circuits;

- □ draw analogies between compound propositions and switching circuits.

Before we can discuss how machines understand us, we must consider how we as humans understand each other.

Human to human communication consists of words strung together in meaningful sentences, and this is where we begin. We shall assume that the words we use and the sentences we construct from them have a commonly understood meaning. Many sentences are capable of being understood as being either true or false, and it is these sentences that form the material of this chapter. In this first chapter we shall see how sentences combine together and how truth or falsity is conveyed by them and how inferences can be made from such combined sentences.

Propositions: the connectives AND, OR and NOT

Information systems and the propositional calculus

An **information system** processes **data** to produce **information**, and each one of us is an information system. Our five senses of sight, hearing, touch, smell and taste are all equipped to absorb data. By relating the data absorbed to its source our brain then extracts information: an arc light is bright, a whisper is quiet, sandpaper is rough, dead mice stink and sugar is sweet.

By combining such extracted information with information already stored in our brain we reason and form conclusions. You look out of the window and see rain, so before you venture outside you pick up your umbrella. This sort of simple reasoning is done automatically with very little conscious effort. However, if your 4-year-old child wanted to go out to play in the puddles your reasoning would be given in greater detail. You would carefully explain that, because it was raining, boots, hat and a raincoat would have to be worn in order to keep dry inside.

Reasoning is governed by a collection of rules called the rules of **logic**, and when we reason we apply these rules to the information that we receive and the information that we store. All information systems, be they biological, manual or computer driven, must operate in accordance with the rules of logic to enable them to function correctly, and so this is where we start – the rules of logic contained within the propositional calculus.

We shall describe the differences between data and information later in the book.

Sentences and syntax

In English a **sentence** is a collection of words that when listed together have meaning. We say that it is semantically correct, where the word **semantics** means **meaning**. For example, the list of words

The balloon is up in the air

has meaning and so is a valid sentence – it is semantically correct. However, the list of words

Air is balloon in the up the

contains the same collection of words but has no meaning – it is not semantically correct and so is not a valid sentence. What makes a sentence valid is not just the words that it contains but also the way the words are put together. For a sentence to be valid the words that it contains must be put together according to the rules of grammatical construction, otherwise known as the rules of **syntax**. This applies not only in the English language but also in computer languages. For example, the list of symbols

x := x + 1

is a valid sentence in the Pascal language which translates into English as

let the new value of x be equal to the old value of x plus 1

Here, in the Pascal language, symbols are used instead of English words, but because the symbols are listed together correctly according to the Pascal syntax they form a valid sentence. On the other hand, the list of symbols

$x + 1 := x$

is not a valid sentence in Pascal because it does not obey the Pascal syntax – it does not obey the rules that regulate the construction of sentences in that language.

Conclusion: A sentence is a collection of words or symbols that are listed together correctly according to an appropriate syntax.

Propositions

Having established what a valid sentence is we now consider a special type of sentence called a **proposition**. A proposition is a **declarative** sentence that declares or **proposes** a single fact and is either true or false. For example, the sentence

Bluebells are blue

is declarative and true, whereas the sentence

Bluebells are yellow

is declarative and false – they are both, however, valid propositions. Each sentence proposes a single fact that is immediately seen to be true or false. We say that a proposition has a **truth value** of true or false.

Not all sentences are declarative and so are not capable of being either true or false. For example, the sentence

What is the time?

is a question and does not declare a fact. Questions are neither true nor false. Because this sentence can be neither true nor false it is not a proposition.

Most sentences are capable of being written in different ways. For example, the sentence

Two plus two equals four

or

Two and two make four

can also be written using arithmetic symbols as

$$2 + 2 = 4$$

All three forms are equally valid because they each obey their appropriate syntax and have the same meaning.

Propositions and instructions
In a computer program a statement of the form

count := count + 1

is an instruction to increase the value of the variable **count** by 1. An instruction cannot be assigned a truth value and so instructions are not propositions. However, if, in a section of pseudocode, the following statement appeared:

count' = count + 1

this would be taken to be read as

the new value of count becomes the old value of count plus 1

and as such is a proposition. The difference between these two interpretations is a very fine distinction but it must be made nonetheless. Program instructions are not propositions but pseudocode statements are propositions. This is despite the fact that both may look very similar.

Conclusion: A proposition is a sentence that declares a fact that is either true or false.

Worked example 1.1

Which of the following are valid propositions?
(a) The day is sunny.
(b) Who is Sylvia?
(c) Up the Rovers!
(d) $4 > 5$
(e) sum := sum − 2
(f) customerNumber' = customerNumber + 1

Solution:

(a) This is a valid proposition because it declares a fact that is either true or false.

(b) This is not a valid proposition because it is a question and cannot be either true or false.

(c) This is not a valid proposition. It is an exclamation and is neither true nor false.

(d) This sentence is false and therefore is a valid proposition.

(e) This may be a valid sentence but it is not necessarily a valid proposition. In some computer languages it is a valid instruction while in others it is not. Even if it is a valid instruction in some language, if it cannot be assigned a truth value then it will not be a valid proposition.

(f) This is a valid pseudocode statement and as such it can be assigned a truth value and hence is a proposition.

Exercise

1.1 Which of the following are valid propositions?

(a) The night was cold.

(b) When shall I go?

(c) Out damned spot.

(d) $18*3 > 50$

(e) stock := stock + delivery

(f) stock$'$ = stock + delivery

Logical connectives

A **compound** proposition is a proposition built up from a collection of simple propositions. For example, the compound proposition

It is lunchtime and I am hungry

is constructed from the simple proposition 'It is lunchtime' and the simple proposition 'I am hungry'. The word **and** that joins them together is called a **logical connective**. Logic is the study of meaning and the compound proposition formed by connecting two simple propositions has a meaning that depends upon the individual meanings of each of the two simple propositions.

There are a number of logical connectives, and in this chapter we shall look at three of them – AND, OR and NOT.

The connective AND

The compound of two propositions connected by AND is called the **conjunction** of the propositions. For example, the conjunction of the proposition

4 is greater than 3

and the proposition

4 is less than 5

is the compound proposition

4 is greater than 3 AND 4 is less than 5

We can rephrase this compound proposition without affecting its meaning:

4 is greater than 3 AND less than 5

but be careful. In isolation, the phrase

less than 5

is no longer a valid proposition. When separating a compound proposition into its component propositions it may be necessary to rephrase the compound proposition.

We can also rewrite these propositions in symbolic form:

$4 > 3$, $4 < 5$ and the compound proposition $3 < 4 < 5$

Notice that $4 > 3$, 4 is greater than 3, means the same as $3 < 4$, 3 is less than 4.

Conclusion: Two propositions can be joined together by using AND to form a single proposition called the conjunction of the two original propositions.

Truth values

The truth value of a compound proposition depends upon the truth values of its component simple propositions. For example, the compound proposition

I am happy AND the sun is shining

is true only if I am happy and the sun is shining. If either the sun is not shining or I am not happy (or both) then the compound proposition is false.

The English use of **both** in combination with **and** is superfluous in logic.

Conclusion: The compound proposition consisting of two simple propositions joined together by AND is true only when both component propositions are true.

Worked example 1.2

Connect the following pairs of propositions using the connective AND and rephrase the conjunction into more natural English or more natural symbolism:
(a) Peter is tall, Jane is tall.
(b) Peter is short, Jane is tall.
(c) $5 < 6$, 6 is less than 8
(d) stock $<$ reOrderLevel, reOrderFlag is ON
(e) stock $<$ reOrderLevel, reOrderFlag is OFF

Solution:
(a) Peter is tall AND Jane is tall: both Peter and Jane are tall.
(b) Peter is short AND Jane is tall: Peter is short but Jane is tall.
(c) $5 < 6$ AND $6 < 8$: $5 < 6 < 8$
(d) stock $<$ reOrderLevel AND reOrderFlag is ON: the stock is less than the reorder level and the reorder flag is ON
(e) stock $<$ reOrderLevel AND reOrderFlag is OFF: the stock is less than the reorder level but the reorder flag is OFF
Note: The use of 'but' as an alternative to 'and' is common in English. The word 'but', however, is not an alternative to 'and' in the propositional calculus.

Worked example 1.3

Translate each of the following sentences into the conjunction of two propositions.
(a) Both customerNumber and supplierNumber $<$ MAX
(b) Paper is ecologically sound but plastic is cheaper.
(c) $6 < 10 < 15$

Solution:
(a) The customer number is less than MAX AND the supplier number is less than MAX
(b) Paper is ecologically sound AND plastic is cheaper.
(c) $6 < 10$ AND $10 < 15$

Worked example 1.4

When is each of the following compound propositions false?
(a) $5 > 4 > 6$
(b) The vegetables were neither fresh nor clean.

(c) Both ends of the string are frayed.

(d) Red and blue are both colours of the rainbow.

(e) The stock is less than the reorder level but the reorder flag is OFF.

Solution:

(a) Always, because the second component proposition is false.

(b) When the vegetables were either fresh or clean or both.

(c) When at least one end is not frayed.

(d) Never – both component propositions are always true.

(e) Either the stock is not less than the reorder level or the reorder flag is ON or both.

Exercises

1.2 Connect the following pairs of propositions using AND and rephrase the conjunction into more natural English or more natural symbolism:

(a) The screen is orange, there is less stress on the eyes.

(b) Three will come, four will not come.

(c) $-4 < -2, -2 < 3$

(d) revenue = price × quantity, profit = revenue − cost

(e) quantity > quantityLimit, delivery = complete

1.3 Translate each of the following sentences into the conjunction of two propositions:

(a) Tom and Dick are both councillors.

(b) Petrol is expensive but diesel is cheaper.

(c) Both the credit given and the quantity delivered exceeded their limits.

1.4 When is each of the following compound propositions false?

(a) My spreadsheet can be displayed in colour but my monitor is monochrome.

(b) Saturn and Venus are both planets of our sun.

(c) Both Houses of Parliament are currently sitting.

(d) The printer is switched on but it is still not working.

(e) $2 < 7 < 4$

The connective OR

The compound of two propositions connected by OR is called the **disjunction** of the two propositions. For example, the disjunction of the proposition

I am not late

and the proposition

I am not early

is the compound proposition

I am not late OR I am not early

We can rephrase this compound proposition without affecting its meaning:

either I am not late or I am not early

The English use of **either** in combination with **or** is superfluous in logic.

The use of the word 'either' in English does not usually permit both propositions to be true simultaneously. It is called an **exclusive or** because in permitting the truth of one proposition it automatically excludes the truth of the other. In logic, however, we permit both component propositions of the compound proposition to be true simultaneously – it is an **inclusive or**. For example

I am not late OR I am not early

can permit the meaning **I am on time** – both not late and not early – both component propositions are then true. In other words the disjunction has another component, **or both**, which is taken for granted.

Conclusion: Two propositions can be joined together by using OR to form a single proposition called the disjunction of the two original propositions. This OR is an inclusive OR and permits both propositions to be simultaneously true.

Truth values
The truth value of a compound proposition depends upon the truth values of its component simple propositions. For example, the compound proposition

It is summer or it is autumn

is false only if it is neither summer nor autumn, that is if it is winter or spring.

Conclusion: The compound proposition consisting of two simple propositions joined together by OR is false only when both component propositions are false.

Worked example 1.5

Connect the following pairs of propositions using OR and rephrase the disjunction into more natural English or more natural symbolism:
(a) Your pain is real, your pain is imaginary.
(b) He is cunning, he is stupid.
(c) You wish to eat, you do not wish to eat.
(d) Your credit exceeds its limit, the quantity exceeds the maximum permitted.

Solution:
(a) Your pain is real OR your pain is imaginary: your pain is either real or imaginary.
(b) He is cunning OR he is stupid: he is either cunning or stupid.
(c) You wish to eat OR you do not wish to eat: you either do or do not wish to eat.
(d) Your credit exceeds its limit OR the quantity exceeds the maximum permitted: either your credit exceeds its limit or the quantity exceeds the maximum permitted.

Worked example 1.6

Translate each of the following compound propositions into the disjunction of two propositions:
(a) Either he or she will come.
(b) The letter will arrive today or tomorrow.
(c) Either the delivery has been made or the stock is below the reorder level.

Solution:
(a) He will come OR she will come.
(b) The letter will arrive today OR the letter will arrive tomorrow.
(c) The delivery has been made OR the stock is below the reorder level.

Worked example 1.7

When is each of the following compound propositions false?
(a) Her pet is a dog or a cat.
(b) Today is Monday or it is 14 January.
(c) The delivery has been made or the stock is under the reorder level.
(d) Either you wish to eat or you do not.

Solution:
(a) Her pet is neither a dog nor a cat.
(b) Today is neither Monday nor 14 January.
(c) The delivery has not been made and the stock is at or above the reorder level.
(d) Never.

Exercises

1.5 Connect the following pairs of propositions using OR and rephrase the conjunction into more natural English or more natural symbolism:
   (a) It is time to go, my watch is fast.
   (b) It has been raining, it was a heavy dew.
   (c) You are reading the book, you returned the book to the library.

1.6 Translate each of the following compound propositions into the disjunction of two propositions:
   (a) Either he leaves or I do.
   (b) To some people that was either good or bad news.
   (c) The answer is yes or no.

1.7 When is each of the following compound propositions false?
   (a) Either the weather brightens up or we do not have our picnic.
   (b) Either your credit card has expired or you have no further credit.
   (c) He will either answer the phone when it rings or he is not in.
   (d) The answer is yes or no.

The connective NOT
The third of the connectives to be considered in this chapter is NOT. This connective differs from the other two in that it acts on a single proposition to form the negation of the proposition. For example, the negation of

   I am happy

is the proposition

   I am NOT happy

which has the **opposite** meaning. In other words, if a proposition is false then its negation is true and if the proposition is true its negation is false.

NOT acts on a single proposition so to call it a connective is not strictly correct. However, the use of the word connective is customary.

*Conclusion: A proposition can be converted into a proposition with the opposite meaning by using NOT to form the **negation** of the original proposition. True propositions are negated to false propositions and false propositions are negated to true propositions.*

Worked example 1.8

Negate the following propositions using NOT and rephrase the negation into more natural English or more natural symbolism:
(a) He is married.
(b) $7 < 10$
(c) π is an irrational number.
(d) salesThisMonth < salesLastMonth

Solution:
(a) He is NOT married: he isn't married.
(b) 7 is NOT less than 10: $7 \geq 10$
(c) π is NOT an irrational number: π is a rational number.
(d) salesThisMonth NOT < salesLastMonth: salesThisMonth $\geq$ salesLast Month

Worked example 1.9

Translate each of the following compound propositions into two connected, negated propositions:
(a) Neither he nor she is needed.
(b) The letter won't arrive today or tomorrow.
(c) Tomorrow is neither Monday nor Tuesday.

Solution:
(a) He is NOT needed AND she is NOT needed.
(b) The letter will NOT arrive today AND the letter will NOT arrive tomorrow.
(c) Tomorrow is NOT Monday AND tomorrow is NOT Tuesday.

Exercises

1.8 Negate the following propositions using NOT and rephrase the negation into more natural English or more natural symbolism:
(a) The animal is not tame.
(b) Bill is a not a non-member of the club.
(c) The letter c is often used to denote a constant.

1.9 Translate each of the following compound propositions into two connected, negated propositions:
(a) Neither 3 red nor 10 blue is sufficient.
(b) They are neither wide nor narrow.
(c) It was the best of times, it was the worst of times.

Propositions and switching circuits

Propositions in general

If we were to continue solely to use English for the format of propositions, then large compound propositions would very soon become too cumbersome to handle. Also, the problem of considering propositions in the form of specific sentences such as

Granny Smith apples are green

is that any conclusions drawn are specific to the example in question. To be able to draw more general conclusions about propositions we need to be more general in our descriptions. To this end we shall use symbols for both the propositions and their connectives.

Simple propositions are denoted by lower-case letters:

p: the cat sat on the mat
q: the dog sat on the floor

Here, the symbol p stands for 'the cat sat on the mat' and the symbol q stands for 'the dog sat on the floor'. Connectives are symbolized as follows:

AND: The symbol $\wedge$ will be used to denote AND. For example

$p \wedge q$ = the cat sat on the mat AND the dog sat on the floor

OR: The symbol $\vee$ will be used to denote OR. For example

$p \vee q$ = the cat sat on the mat OR the dog sat on the floor

NOT: Negation is represented by the symbol $\neg$. For example

$\neg p$ = the cat did NOT sit on the mat

Conclusion: Symbols can be used to represent both propositions and their connectives.

Worked example 1.10

Translate into symbolic form:
(a) June is a warm month and July is a warm month.
(b) Both June and July are warm months.
(c) Either June is a warm month or July is a warm month.
(d) Neither June nor July is a warm month.

(e)   June is a warm month but July is not a warm month.
(f)   June and July are not warm months.
(g)   Either the stock level is greater than 100 or the order request is reissued and the request is recorded.

Solution: Let p = 'June is a warm month' and q = 'July is a warm month'.

(a) $p \wedge q$
(b) $p \wedge q$
(c) $p \vee q$
(d) $\neg p \wedge \neg q$
(e) $p \wedge \neg q$
(f) $\neg p \wedge \neg q$
(g) $a \vee (b \wedge c)$, where a = 'stock level > 100', b = 'order request reissued' and c = 'request is recorded'. There is a possibility of ambiguity here. The sentence could be translated to read as $(a \vee b) \wedge c$.

Worked example 1.11

Translate into English where:
 (i)   p = 'Yesterday was my birthday', q = 'Tomorrow is your birthday'
(ii)   p = 'customerNumber < 150', q = 'supplierNumber > 10'

(a) $p \wedge q$
(b) $p \wedge \neg q$
(c) $\neg p \wedge \neg q$
(d) $\neg p \vee \neg q$
(e) $\neg(p \wedge q)$

Solution:
(ia)   Yesterday was my birthday and tomorrow is your birthday.
(ib)   Yesterday was my birthday but tomorrow is not your birthday.
(ic)   Yesterday was not my birthday and tomorrow is not your birthday.
(id)   Either yesterday was not my birthday or tomorrow is not your birthday.
(ic)   It is not correct that both yesterday was my birthday and tomorrow is your birthday.

(iia) The customerNumber < 150 and the supplierNumber > 10
(iib) The customerNumber < 150 and the supplierNumber ≤ 10
(iic) The customerNumber ≥ 150 and the supplierNumber ≤ 10
(iid) Either the customerNumber ≥ 150 or the supplierNumber ≤ 10
(iie) Either the customerNumber ≥ 150 or the supplierNumber ≤ 10

Note: These answers are, of course, not the only solutions. English is a language in which the same meaning can be transmitted in a variety of forms.

Exercises

1.10 Translate into symbolic form:
  (a) Apples are either green or sweet.
  (b) Apples are neither green nor sweet.
  (c) Apples are both green and sweet.
  (d) Apples are not green and sweet.
  (e) Apples are green but not sweet.
  (f) Apples are green and apples are sweet.
  (g) Either the customer is on record or the customer is new and must be recorded.

1.11 Translate into English:
  (a) $p \vee q$
  (b) $p \wedge \neg q$
  (c) $\neg p \vee \neg q$
  (d) $\neg p \vee q$
  (e) $\neg(p \vee q)$
  where
  (i) p = It is August, q = Today is Thursday
  (ii) p = Payment has been made, q = Delivery is complete

Truth values

We have stated that a proposition is a declarative statement and that it is either true or false. For example, consider the proposition

  It is raining

Look out of the window. If it is raining then this proposition is true, otherwise it is false. In the language of logic we say that a proposition assumes one of two **truth values**: truth, symbolized by **T**, and falsehood, symbolized by **F**. Some propositions may assume either truth value as in the proposition

  Today is Monday

Other propositions may be able to assume only one fixed truth value. For example, the proposition

  $1 + 1 = 2$

can possess only the truth value T – we call such a proposition a **tautology**. Again, the proposition

$$1 + 1 = 3$$

can only possess the truth value F – such a proposition is called it a **contradiction**.

Truth tables

A truth table is a device that permits the truth value of a compound proposition to be found when the truth values of the component propositions are given. Now we are using symbols to represent our propositions we can no longer state categorically that a given proposition is true or it is false. Instead we must allow for either truth value to apply, and this is achieved within the truth table.

For example, consider the compound proposition

I am called James and I am 17 years old

This compound proposition is true only if I am both called James and am 17 years old. If I am a 17-year-old John then the compound proposition is as false as if I were a 94-year-old James.

In other words, this compound proposition assumes the truth value T only when both component propositions assume the truth value T. We can summarize this result in a **truth table**.

Let p stand for 'I am James' and q stand for 'I am 17 years old' so that $p \wedge q$ stands for the conjunction of these two propositions. The truth table is then:

p	q	$p \wedge q$
F	F	**F**
F	T	**F**
T	F	**F**
T	T	**T**

This truth table will result from the conjunction of any two propositions, regardless of their content. We call it the truth table for the connective AND.

AND

The truth table for the conjunction of the two propositions p and q is:

p	q	$p \wedge q$
F	F	F
Г	T	Г
T	F	F
T	T	T

Notice the distinction between the use of the word **tautology** in mathematics and its definition in the *Oxford English Dictionary*, where it is defined to mean saying the same thing twice over in different words. Words used in mathematics do occasionally have meanings different from their dictionary definitions, and care must always be taken to ensure that you understand the meanings of words when they are used in a mathematical context.

The only time a conjunction has truth value T is when both component propositions are true.

OR

The truth table for the disjunction of the two propositions p and q is

p	q	p ∨ q
F	F	F
F	T	T
T	F	T
T	T	T

The only time a disjunction has truth value F is when both component propositions are false. For example

   I am called James or I am 17 years old

is false only if I am neither called James nor am 17 years old. If I am a 17-year-old John then the compound proposition is as true as if I were a 94-year-old James.

NOT

The truth table for the negation of proposition p is very simple. It is

p	¬p
F	T
T	F

Truth tables can be constructed for finding the truth values of compound propositions that are more complicated than those we have just considered. For example, the truth values of the compound proposition ¬p ∧ (q ∨ r) are found from a truth table as follows.

The variables p, q and r are listed alongside the expression involving the variables. Take care to leave sufficient space between the symbols to insert the truth values without cramping the overall effect. First we enter the eight possible combinations of truth values as shown in the first table.

p	q	r	¬p	∧	(q	∨	r)	p	q	r	¬p	∧	(q	∨	r)
F	F	F						F	F	F	T				
F	F	T						F	F	T	T				
F	T	F						F	T	F	T				
F	T	T						F	T	T	T				
T	F	F						T	F	F	F				
T	F	T						T	F	T	F				
T	T	F						T	T	F	F				
T	T	T						T	T	T	F				
1	**2**	**3**						**1**	**2**	**3**	**4**				

The truth values for ¬p are then entered. Notice that the numbers in the bottom row of the table give the order in which the table is being completed. You will find this a useful habit to acquire to avoid confusion when you are completing large tables.

In the third table the truth values for q ∨ r are entered:

p	q	r	¬p	∧	(q	∨	r)
F	F	F	T		F	F	F
F	F	T	T		F	T	T
F	T	F	T		T	T	F
F	T	T	T		T	T	T
T	F	F	F		F	F	F
T	F	T	F		F	T	T
T	T	F	F		T	T	F
T	T	T	F		T	T	T
1	2	3	4		5	7	6

p	q	r	¬p	∧	(q	∨	r)
F	F	F	T	F	F	F	F
F	F	T	T	T	F	T	T
F	T	F	T	T	T	T	F
F	T	T	T	T	T	T	T
T	F	F	F	F	F	F	F
T	F	T	F	F	F	T	T
T	T	F	F	F	T	T	F
T	T	T	F	F	T	T	T
1	2	3	4	8	5	7	6

Finally, the truth values for the compound expression are found in column 8 of the fourth table by comparing the truth values between columns 4 and 7.

Conclusion: Every proposition assumes a truth value of T or F. The truth value of a compound proposition is dictated by the truth values of its component propositions.

Worked example 1.12

Construct the truth tables of each of the following compound propositions:
(a) (¬p) ∧ q
(b) p ∨ ¬q
(c) (p ∧ q) ∨ (¬p ∨ q)

Solution:

(a)

p	q	¬p	∧	q
F	F	T	F	F
F	T	T	T	T
T	F	F	F	F
T	T	F	F	T
1	2	3	5	4

(b)

p	q	p	∨	¬q
F	F	F	**T**	T
F	T	F	**F**	F
T	F	T	**T**	T
T	T	T	**T**	F

The order in which to complete the columns in (b) and (c) has been left for you to work out.

(c)

p	q	(p	∧	q)	∨	(¬p	∨	q)
F	F	F	F	F	**T**	T	T	F
F	T	F	F	T	**T**	T	T	T
T	F	T	F	F	**F**	F	F	F
T	T	T	T	T	**T**	F	T	T

Exercise

1.12 Construct the truth tables of each of the following compound propositions:

(a) $(\neg p) \wedge p \wedge q$

(b) $(q \vee \neg q) \wedge p$

(c) $((\neg\neg p) \wedge (\neg q)) \vee ((\neg p) \vee q)$

Switches

The heart of a computer – the central processing unit – contains circuitry that involves a very large number of devices that are best described as switches. This is because they are either ON, so permitting electric current to flow across them, or they are OFF, thereby inhibiting the flow of electric current.

Figure 1.1

In Figure 1.1 the switch is denoted by the capital letter A. It is always drawn in the open position – OFF – but its state can be either ON or OFF. If the switch is ON then it is assigned the state value 1:

$A := 1$

That is current can flow through the switch. If the switch is OFF then it is assigned state value 0:

$A := 0$

That is current cannot flow through the switch.

Switching circuits

The circuits in the central processing unit contain many of these switches connected together to form what are called **switching circuits**. Within a switching circuit any pair of connected switches can be joined together in only one of two ways.

Series connection

Two switches can be joined end to end to form a series connection as shown in Figure 1.2.

Figure 1.2

From the diagram it can be seen that current will flow through the two connected switches only if both switches are ON, that is both are assigned a state value of 1. If we use ∧ to denote a series connection then the state value of this compound switch can be described using the following table:

A	B	A ∧ B
0	0	**0**
0	1	**0**
1	0	**0**
1	1	**1**

Note the similarity between this table and the truth table for the logical connective AND when considering propositions. If we were to replace 0 by F and 1 by T then the two tables would be identical.

A	B	A ∧ B
F	F	F
F	T	F
T	F	F
T	T	T

Parallel connection

Two switches can be joined to form a parallel connection as shown in Figure 1.3.

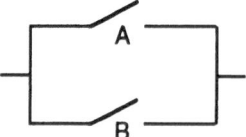

Figure 1.3

From the diagram it can be seen that current will not flow through the two connected switches only if both switches are OFF – that is both are assigned a state value of 0. If we use ∨ to denote a parallel connection then the state value of this compound switch can be described using the following table:

A	B	A ∨ B
0	0	**0**
0	1	**1**
1	0	**1**
1	1	**1**

Note the similarity between this table and the truth table for the logical connective OR when considering propositions.

Notice also that the listing of possible combinations of state values in these two tables is in the form of a binary number of increasing value, 00, 01, 10 and 11.

Negated switch

If A represents a switch then ¬A represents the negated version of the switch. By negation is meant that if A is ON then ¬A is OFF and if A is OFF then ¬A is ON. Again, we can construct a table for ¬ as follows:

A	¬A
0	**1**
1	**0**

Again, note the similarity between this table and the truth table for the negation of a proposition.

Worked example 1.13

Construct tables for each of the switching circuits shown in (a) Figure 1.4, (b) Figure 1.5, (c) Figure 1.6 and (d) Figure 1.7.

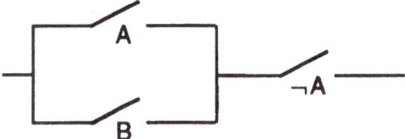

Figure 1.4

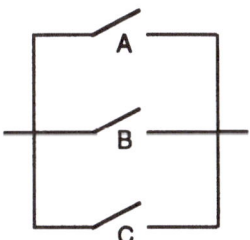

Figure 1.5

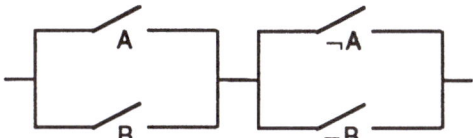

Figure 1.6

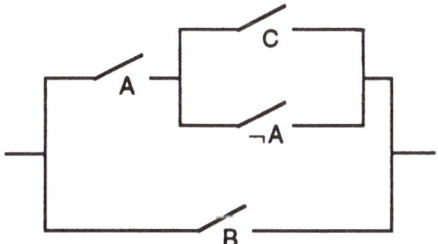

Figure 1.7

Solution:

(a) The circuit is described by the following:

$(A \lor B) \land \neg A$

and has the following truth table:

A	B	(A	∨	B)	∧	¬A
0	0	0	0	0	**0**	1
0	1	0	1	1	**1**	1
1	0	1	1	0	**0**	0
1	1	1	1	1	**0**	0
1	**2**	**3**	**6**	**4**	**7**	**5**

(b) The circuit is described by the following:

A ∨ B ∨ C

and has the following truth table:

A	B	C	(A	∨	B)	∨	C
0	0	0	0	0	0	**0**	0
0	0	1	0	0	0	**1**	1
0	1	0	0	1	1	**1**	0
0	1	1	0	1	1	**1**	1
1	0	0	1	1	0	**1**	0
1	0	1	1	1	0	**1**	1
1	1	0	1	1	1	**1**	0
1	1	1	1	1	1	**1**	1
1	**2**	**3**	**4**	**7**	**5**	**8**	**6**

(c) The circuit is described by the following:

(A ∨ B) ∧ (¬A ∨ ¬B)

and has the following truth table:

A	B	(A	∨	B)	∧	(¬A	∨	¬B)
0	0	0	0	0	**0**	1	1	1
0	1	0	1	1	**1**	1	1	0
1	0	1	1	0	**1**	0	1	1
1	1	1	1	1	**0**	0	0	0
1	**2**	**3**	**7**	**4**	**9**	**5**	**8**	**6**

(d) The circuit is described by the following:

[A ∧ (C ∨ ¬A)] ∨ B

and has the following truth table:

A	B	C	[A	∧	(C	∨	¬A)]	∨	B
0	0	0	0	0	0	1	1	**0**	0
0	0	1	0	0	1	1	1	**0**	0
0	1	0	0	0	0	1	1	**1**	1
0	1	1	0	0	1	1	1	**1**	1
1	0	0	1	0	0	0	0	**0**	0
1	0	1	1	1	1	1	0	**1**	0
1	1	0	1	0	0	0	0	**1**	1
1	1	1	1	1	1	1	0	**1**	1
1	**2**	**3**	**4**	**9**	**5**	**8**	**6**	**10**	**7**

Exercise

Construct tables for each of the switching circuits shown in (a) Figure 1.8, (b) Figure 1.9 and (c) Figure 1.10.

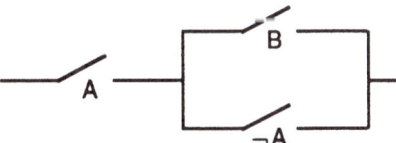

Figure 1.8

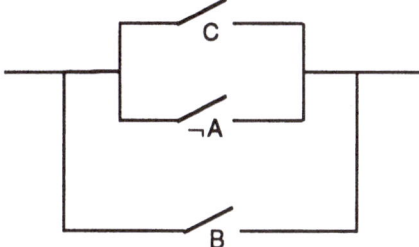

Figure 1.9

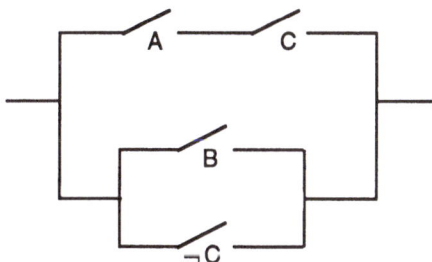

Figure 1.10

Propositions and switches

The similarity between the description of switches and the description of propositions is evident. A switch is a two-state device: it is either ON or OFF. A proposition is a two-state item – it is either TRUE or FALSE. The SERIES connection of two switches is completely analogous to two propositions connected by AND. The PARALLEL connection of two switches is completely analogous to two propositions connected by OR. Switches and propositions can both be negated. In other words, the description of switches is immediately analogous to the description of propositions where state values 0 and 1 correspond to truth values T and F respectively. Herein lies the heart of the connectivity between the propositional calculus and the behaviour of the computer. To maintain this connectivity explicitly we shall use 0 and 1 to represent the truth values F and T respectively.

Binary form of truth tables for propositions

In summary, we rewrite the truth tables for the logical connectives AND, OR and for the negation NOT using the binary digits 0 and 1 as follows:

p	q	p $\wedge$ q
0	0	0
0	1	0
1	0	0
1	1	1

p	q	p $\vee$ q
0	0	0
0	1	1
1	0	1
1	1	1

p	$\neg$q
0	1
1	0

Worked example 1.14

Construct truth tables using the binary digits 0 and 1 for each of the following compound propositions:

(a) $(p \vee q) \wedge \neg p$
(b) $(p \vee q) \wedge (\neg p \vee \neg q)$
(c) $p \vee q \vee r$
(d) $(p \wedge (q \vee \neg p)) \vee r$

Solution:

(a)
p	q	(p	$\vee$	q)	$\wedge$	$\neg$p
0	0	0	0	0	0	1
0	1	0	1	1	1	1
1	0	1	1	0	0	0
1	1	1	1	1	0	0

(b)

p	q	(p	∨	q)	∧	(¬p	∨	¬q)
0	0	0	0	0	**0**	1	1	1
0	1	0	1	1	**1**	1	1	0
1	0	1	1	0	**1**	0	1	1
1	1	1	1	1	**0**	0	0	0

(c)

p	q	r	p	∨	q	∨	r
0	0	0	0	0	0	**0**	0
0	0	1	0	0	0	**1**	1
0	1	0	0	1	1	**1**	0
0	1	1	0	1	1	**1**	1
1	0	0	1	1	0	**1**	0
1	0	1	1	1	0	**1**	1
1	1	0	1	1	1	**1**	0
1	1	1	1	1	1	**1**	1
1	**2**	**3**	**4**	**7**	**5**	**8**	**6**

(d)

p	q	r	(p	∧	(q	∨	¬p))	∨	r
0	0	0	0	0	0	1	1	**0**	0
0	0	1	0	0	0	1	1	**1**	1
0	1	0	0	0	1	1	1	**0**	0
0	1	1	0	0	1	1	1	**1**	1
1	0	0	1	0	0	0	0	**0**	0
1	0	1	1	0	0	0	0	**1**	1
1	1	0	1	1	1	1	0	**1**	0
1	1	1	1	1	1	1	0	**1**	1
1	**2**	**3**	**4**	**9**	**5**	**8**	**6**	**10**	**7**

Exercise

Construct truth tables using the binary digits 0 and 1 for each of the following compound propositions:

(a) $p \vee (q \wedge \neg p)$

(b) $p \wedge (q \vee \neg p)$

(c) $\neg p \vee q \vee r$

(d) $(p \wedge q) \vee (r \vee \neg q)$

Chapter 2

Truth and consequences

OBJECTIVES

When you have completed this chapter you will be able to:

- recognize the sequence of condition followed by consequence as an implication;

- explain the idea of a sufficient condition for a subsequent consequence;

- construct the truth table for the implication connective;

- explain the meanings of the biconditional and necessity;

- reproduce the biconditional truth table;

- demonstrate the meaning of logical equivalence;

- prove or disprove two compound propositions to be logically equivalent;

- reduce compound propositions to simpler forms by using the rules of logic;

- construct equivalent propositions from the table truth of a compound proposition;

- demorganize a compound proposition;

- reduce a compound proposition to a simpler equivalent form.

A computer is a system that accepts input, processes the input and outputs the result of its processing. The processing of the input is controlled by programs – collections of commands – that instruct the computer to perform certain tasks. To ensure that the program causes the computer to perform the operations that we require it to perform the program must be correctly constructed. One of the ways in which we can be sure that the program is correctly constructed is to ensure that the commands are obeyed – executed – in the correct sequence.

There are three methods of controlling the sequence of execution of a program's instructions. The first is the straightforward **sequential** control in which each statement in the program is executed after its immediate predecessor in the coding has been executed. The second method of sequence control is that which causes a group of commands to be executed repetitively – the **loop** control. The third method of sequence control is **branching**, in which execution is directed to the next command according to the satisfaction of a condition.

The purpose of this chapter is to demonstrate how logic plays its part in the construction of the last two methods of sequence control

The implication connective

Implication

If I walk in the rain then I shall get wet

When humans communicate with each other they convey not only factual information but also conditional and consequential information, and the above sentence is an example of this. Consequences can be implied from previously given conditional information and are a result of the reasoning process. All conditions coupled with consequences – or **implications** – are compound propositions of the form:

IF <proposition 1> THEN <proposition 2>

and in the above example

proposition 1 := I walk in the rain

is the **conditional** proposition and

proposition 2 := I shall get wet

is the **consequent** proposition.

Within the system of logic implication is a logical connective that joins two propositions together and is represented in symbolic form as an arrow →. Hence, the above implication can be written symbolically as

<proposition 1> → <proposition 2>

or, more generally

p → q

which is read as proposition p **implies** proposition q.

Conclusion: An implication is a compound proposition where a conditional proposition implies a consequent proposition. Every implication can be posed in the form IF... THEN...

Worked example 2.1

Write each of the following in symbolic form:
(a) If it is ten o'clock then I am late for my appointment.
(b) When you hear a cuckoo you can be sure it is spring.
(c) I shall help you if you wish.
(d) If DayNumber = 1 then Day = Monday

Solution: The symbolic form is $p \rightarrow q$ where:
(a) p = 'It is ten o'clock': q = 'I am late for my appointment'
(b) p = 'You hear a cuckoo': q = 'You can be sure it is spring'
(c) p = 'You wish me to help you': q = 'I shall help you'
(d) p = 'DayNumber = 1': q = 'Day = Monday'

Worked example 2.2

Translate each of the following into English:
(a) The radio is on $\rightarrow$ I am home
(b) $p \rightarrow \neg q$ where p = 'It is Sunday' and q = 'I go to work'
(c) $\neg p \rightarrow \neg q$ where:
      p = 'The printer is switched on'
      q = 'The document will be printed'
(d) $\neg(p \vee q) \rightarrow (\neg p \wedge \neg q)$ where:
      p = 'The snooker ball is red'
      q = 'The snooker ball is blue'

Solution:
(a) If the radio is on then I am home.
(b) If it is Sunday then I do not go to work.
(c) If the printer is switched off then the document will not be printed.
(d) If the snooker ball is neither red nor blue then the snooker ball is not red and it is not blue.

Exercises

2.1 Write each of the following in symbolic form:
      (a) If it is still daylight at midnight then we are above the Arctic Circle.

    (b)   When the ink runs out the pen will not write.

    (c)   You can take the motorway route if you do not mind the heavy traffic.

    (d)   If DepartmentNumber = 4 then Department = Sales

2.2 Translate each of the following into English:

    (a)   The book is on my chair → I am reading the book

    (b)   p → q where
         p = 'The sun is shining'
         q = 'I carry an umbrella'

    (c) $\neg p \to \neg q$ where:
         p = 'The lecturer is on time'
         q = 'There will be enough time to cover the material'

    (d) $\neg(p \wedge q) \to (\neg p \vee \neg q)$ where:
         p = 'It is time'
         q = 'It is appropriate'

The implication truth table

Just as the logical connectives AND and OR and the negation NOT have truth tables, so has the logical connective IMPLICATION. However, the construction of the truth table for implication is not as straightforward as the earlier tables.

Consider the implication:

If you are a good child then I shall reward you

Bribery! If the child is good and I reward the child then all is well. If the child is bad and I decide not to reward the child then there can be no complaints. Even if I decide to reward the bad child I have still broken no promise. If, however, the child is good and I decide not to reward the child then look out – tantrums could be on the horizon. In this case I have broken a promise. I have been seen to talk falsely. This state of affairs can be summarized in the following table:

Condition	Consequence	Outcome
Child NOT good	NO reward	All is well
Child NOT good	Reward	All is well
Child good	NO reward	All is NOT well
Child good	Reward	All is well

This table can be translated into a truth table for the logical connective IMPLICATION if we use 1 to designate Child good, Reward and All is well and we use 0 to designate their opposites:

p	q	p → q
0	0	**1**
0	1	**1**
1	0	**0**
1	1	**1**

From this truth table we can see that the only time an implication is false is when a truth implies a falsehood.

Sufficiency

Make sure that you really do appreciate the meaning of sufficiency. The meaning of necessity will be made clearer later in this chapter.

Within an implication the conditional proposition is **sufficient** for the consequent proposition, that is the consequence occurs if the condition is satisfied. For example, a child being good is a sufficient condition for a reward to be given – but it is not necessary. A reward can be given even if the child is not good. The consequence can occur even if the condition is not met.

Conclusion: The conditional proposition in an implication is sufficient for the consequence and the only time an implication is false is when a true condition implies a false consequence.

Worked example 2.3

Construct truth tables for each of the following:
(a) $p \rightarrow \neg q$
(b) $(p \vee q) \rightarrow (q \vee p)$
(c) $(p \wedge q) \rightarrow (p \vee q)$
(d) $\neg(p \vee q) \rightarrow (\neg p \wedge \neg q)$

Solution:

(a)

p	q	p	→	¬q
0	0	0	**1**	1
0	1	0	**1**	0
1	0	1	**1**	1
1	1	1	**0**	0

(b)

p	q	(p	∨	q)	→	(q	∨	p)
0	0	0	0	0	**1**	0	0	0
0	1	0	1	1	**1**	1	1	0
1	0	1	1	0	**1**	0	1	1
1	1	1	1	1	**1**	1	1	1

(c)

p	q	(p	∧	q)	→	(p	∨	q)
0	0	0	0	0	**1**	0	0	0
0	1	0	0	1	**1**	0	1	1
1	0	1	0	0	**1**	1	1	0
1	1	1	1	1	**1**	1	1	1

(d)	**p**	**q**	¬	(**p**	∨	**q**)	→	(¬**p**	∧	¬**q**)
	0	0	1	0	0	0	**1**	1	1	1
	0	1	0	0	1	1	**1**	1	0	0
	1	0	0	1	1	0	**1**	0	0	1
	1	1	0	1	1	1	**1**	0	0	0

Notice that we have omitted the numbering on the columns as the order in which the table is completed should by now be clear. Notice also that the last three of these implications are tautologies.

Exercise

2.3 Construct truth tables for each of the following:
  (a) ¬p → q
  (b) (p ∧ q) → (q ∧ p)
  (c) (p ∨ q) → (p ∧ q)
  (d) ¬(p ∧ q) → (¬p ∨ ¬q)

Boolean variables

Within a computer program it is possible to define variables, a variable being a named item that can assume different values of a declared type. For example, if, in a Pascal program, the named item **Number** is declared to be an integer it can thenceforth be assigned integer values:

```
Number := 6
```

A Boolean variable is a declared variable that can be assigned either one of two values, TRUE or FALSE. For example, let p be a declared Boolean variable which is later assigned the proposition Number > 5:

```
p := Number > 5
```

If Number = 4 then p has the value FALSE and if Number = 6 then p has the value TRUE.

If ... then ...
The computer program command structure:

```
If <proposition> then <command>
```

is called a branching control structure. If the <proposition> is TRUE then

Boolean variables are so called after George Boole (1815–64), who defined them in his development of Boolean algebra.

the <command> will be executed. If the <proposition> is FALSE then the <command> will not be executed. For example, consider the following Pascal program:

```
program prog01(output);
var     Day : string;

begin
        Day := Monday;
        If Day = Monday then writeln ('Today is Monday.')
end.
```

Here the variable Day is declared to be a string variable and is assigned the string value Monday. In the If... then... command structure the proposition Day = Monday is TRUE. Consequently, the writeln command is executed. If Day had been assigned Tuesday then the proposition Day = Monday would be FALSE and the writeln command would not have been executed.

Conclusion: Within a computer program a Boolean variable is declared to be a named item that can be assigned either of the two values TRUE or FALSE. Within the If <proposition> then <command>.command structure of a computer program, if <proposition> is TRUE then <command> will be executed. If <proposition> is FALSE then <command> will not be executed.

Looping
A further computer program sequence control structure is that which controls **looping**, in which a block of commands are executed repetitively. Two such control structures are given by **repeat** and **while**.

Repeat
The looping control command repeat is of the form:

```
repeat
        <command>
        <command>

        ...
        <command>
until <proposition>
```

Here the block of commands will be executed repetitively as long as <proposition> is FALSE or until <proposition> is TRUE. Exiting from the loop is caused when <proposition> is TRUE. For example, in the following Pascal code:

```
program prog02(input,output);
var       Number : integer;

begin
        Number := 0;
        repeat
          writeln (Number);
          read (Number);
        until Number = -999
end.
```

the variable Number is declared to be an integer variable and is assigned the value 0. The loop is entered at the command repeat. The value is then printed and the new value for Number is input. If this input value is not −999 then the proposition Number = −999 is FALSE and so execution jumps back to the start of the loop. The new value of Number is printed, after which another new value for Number is input. If this time the input is −999 then the proposition Number = −999 is TRUE and execution leaves the loop and the program ends.

While

The looping control command while is of the form:

```
while <proposition> do
begin
            <command>
            <command>
            ...
            <command>
end;
```

Here the block of commands will be executed repetitively as long as <proposition> is TRUE or until the <proposition> is FALSE. Exiting from the loop is caused when <proposition> is FALSE. For example, in the following Pascal code:

```
program prog03(input,output);
var       Number : integer;

begin
        Number := 0;
        while Number <> -999 do
        begin
          writeln (Number);
          read (Number)
        end
end.
```

the variable Number is declared to be an integer variable and is assigned the value 0. The loop is entered at the command while because the proposition Number <> −999 is TRUE. Within the loop the value is then printed and new value for Number is input. Execution returns to the start of the loop, and if the input value is not −999 the proposition Number <> −999 is TRUE and so execution continues through the loop. The new value of Number is printed, after which another new value for Number is input. If this time the input is −999 then the proposition Number <> −999 is FALSE and execution leaves the loop and the program ends.

Conclusion: The repetitive execution of a block of code can be controlled by a looping command control structure. Two such commands are repeat *and* while. Repeat *causes a block of code to be repetitively executed provided a given proposition is FALSE, whereas* while *causes a block of code to be repetitively executed provided a given proposition is TRUE.*

Worked example 2.4

Consider the following Pascal program:

```
program prog 04(input,output);
var      p : boolean;
         Number : integer;

begin
         read (Number);
         p := Number > 5;
         if p then writeln ('Cannot be counted on one hand.');
         if NOT p then writeln ('Can be counted on one hand.')
end.
```

Describe the output from the following program when the input assigns to the variable Number the value:

(a) 3
(b) 5
(c) 7

Solution:
(a) Because Number := 3 then p = Number > 5 is FALSE and NOT p is TRUE. The output is, accordingly, Can be counted on one hand.
(b) Because Number := 5 then p = Number > 5 is FALSE and NOT p is TRUE. The output is, accordingly, Can be counted on one hand.
(c) Because Number := 7 then p = Number > 5 is TRUE. The output is, accordingly, Cannot be counted on one hand.

Worked example 2.5

Consider the following Pascal program:

```
program prog05(input,output);
var     p : boolean;
        Balance, CustomerCredit : real;

begin
        read (Balance);
        read (CustomerCredit);
        p := (Balance > 125) AND (CustomerCredit = 100);
        if p then writeln ('Do not despatch.');
        if NOT p then writeln ('Despatch.')
end.
```

Describe the output from the following program when the input assigns to the variables Balance and CustomerCredit the respective values:
(a) 100, 100
(b) 130, 100
(c) 100, 90
(d) 150, 120

Solution:
(a) Because Balance := 100 and CustomerCredit := 100 then Balance > 125 is FALSE and CustomerCredit = 100 is TRUE. Consequently p is FALSE, being a conjunction of a true proposition and a false proposition. Accordingly NOT p is TRUE thereby causing the output: Despatch.
(b) Because Balance := 130 and CustomerCredit := 100 then Balance > 125 is TRUE and CustomerCredit = 100 is TRUE. Consequently p is TRUE, being a conjunction of two true propositions. Accordingly NOT p is FALSE thereby causing the output: Do not despatch.
(c) Because Balance := 100 and CustomerCredit := 90 then Balance > 125 is FALSE and CustomerCredit = 100 is FALSE. Consequently p is FALSE, being a conjunction of two false propositions. Accordingly NOT p is TRUE thereby causing the output: Despatch.
(d) Because Balance := 150 and CustomerCredit := 120 then Balance > 125 is TRUE and CustomerCredit = 100 is FALSE. Consequently p is FALSE, being a conjunction of a true proposition and a false proposition. Accordingly NOT p is TRUE thereby causing the output: Despatch.

Worked example 2.6

Describe the execution sequence in the following code:

```
program prog06(output);
var     Count : integer;

begin
        Count := 0;
        repeat
          Count := Count + 1;
          writeln (Count);
        until Count = 4
end.
```

Solution: This program will print the numbers 1, 2, 3 and 4 in sequence. After it has output the number 4 the proposition Count = 4 is tested and found to be TRUE. As a result, execution leaves the loop and the program ends.

Worked example 2.7

Describe the execution sequence in the following code:

```
program prog07(output);
var     Count : integer;

begin
        Count := 0;
        while Count <> 4 do
        begin
          Count := Count + 1;
          writeln (Count)
        end
end.
```

Solution: This program will print the numbers 1, 2, 3 and 4 in sequence. After it has output the number 4 the proposition Count <> 4 is tested and found to be FALSE. As a result, execution leaves the loop and the program ends.

Exercises

2.4 Consider the following Pascal program:

```
program prog08(input,output);
var     p : boolean;
        Temperature : integer;
```

```
begin
        read (Temperature);
        p := Temperature > 100;
        if p then writeln ('Water is boiling.');
        if NOT p then writeln ('Water is not boiling.')
end.
```

Describe the output from the following program when the input assigns to the variable Temperature the value:
(a) 75
(b) 100
(c) 120

2.5 Consider the following Pascal program:

```
program prog09(input,output);
var     p : boolean;
           Balance, CustomerCredit : real;

begin
        read (Balance);
        read (CustomerCredit);
        p := (Balance < 75) OR (CustomerCredit > 0);
        if NOT p then writeln ('Do not despatch.');
        if p then writeln ('Despatch.')
end.
```

Describe the output from the following program when the input assigns to the variables Balance and CustomerCredit the respective values:
(a) 50, 0
(b) 90, 0
(c) 45, 100
(d) 250, 50

2.6 Describe the execution sequence in the following code:

```
program prog 10(output);
var     Count : integer;

begin
        Count := 0;
        repeat
          writeln (Count);
           Count := Count + 1;
        until Count = 4
end.
```

2.7 Describe the execution sequence in the following code:

```
program prog 11(output);
var     Count : integer;
```

```
begin
        Count := 0;
        while Count <> 4 do
        begin
          writeln (Count);
          Count := Count + 1
        end
end.
```

The biconditional and necessity

We have seen that in the compound proposition p → q that p is a **sufficient** condition for q. For example

If today is Friday then we have fish for supper

or

Today is Friday → we have fish for supper

However, this compound proposition does not say that if we have fish for supper then it must be Friday – it could be Monday: being Friday is not a **necessary** condition for having fish for supper. For today being Friday to be a necessary condition for having fish for supper then having fish for supper must also be sufficient for today being Friday. That is

If today is Friday then we have fish for supper

and

If we have fish for supper then today is Friday

This can be alternatively phrased using **if and only if** as

The phrase **if and only if** is abbreviated to **iff**.

If and only if today is Friday then we have fish for supper

Here, p is a necessary condition for q:

p → q and q → p **or** (p → q) ∧ (q → p)

Because necessity between two propositions is such a common occurrence and because the logic expression for it is so unwieldy, an alternative notation has been devised for it. This is

$$p \leftrightarrow q$$

where the logical connective $\leftrightarrow$ is referred to as the **biconditional**. Notice that if p is necessary for q then q is also necessary for p.

Truth table

The truth table for the biconditional is found from the following:

p	q	(p	→	q)	∧	(q	→	p)
0	0	0	1	0	**1**	0	1	0
0	1	0	1	1	**0**	1	0	0
1	0	1	0	0	**0**	0	1	1
1	1	1	1	1	**1**	1	1	1

Hence:

p	q	p	↔	q
0	0	0	**1**	0
0	1	0	**0**	1
1	0	1	**0**	0
1	1	1	**1**	1

The biconditional is only true when both p and q have the same truth values.

*Conclusion: The biconditional signified by **if and only if**... represents the necessity of each component proposition for the other.*

Worked example 2.8

Construct truth tables for:
(a) $[(p \wedge q) \vee r] \leftrightarrow r$
(b) $(\neg p \vee q) \leftrightarrow (p \rightarrow q)$

Solution:

(a)

p	q	r	[(p	∧	q)	∨	r]	↔	r
0	0	0	0	0	0	0	0	1	0
0	0	1	0	0	0	1	1	1	1
0	1	0	0	0	1	0	0	1	0
0	1	1	0	0	1	1	1	1	1
1	0	0	1	0	0	0	0	1	0
1	0	1	1	0	0	1	1	1	1
1	1	0	1	1	1	1	0	0	0
1	1	1	1	1	1	1	1	1	1

(b)	p	q	(¬p	∨	q)	↔	(p	→	q)
	0	0	1	1	0	**1**	0	1	0
	0	1	1	1	1	**1**	0	1	1
	1	0	0	0	0	**1**	1	0	0
	1	1	0	1	1	**1**	1	1	1

Notice that this last compound proposition is a tautology – its truth value is 1 regardless of the truth values of its components.

Exercise

2.8 Construct truth tables for:

(a) $[p \wedge (q \vee r)] \leftrightarrow p$

(b) $\neg(p \to q) \leftrightarrow (p \wedge \neg q)$

Equivalence and the rules of logic

Logical equivalence
Let p be the proposition

p = 'This pencil is red'

and q the proposition

q = 'This pencil is sharp'

The conjunction of p and q is then

$p \wedge q$ = 'This pencil is red AND this pencil is sharp'

which means the same as

This pencil is both red and sharp

The negation of this conjunction is then

This pencil is NOT both red and sharp = $\neg(p \wedge q)$

which means the same as

This pencil is NOT red OR this pencil is NOT sharp = $\neg p \vee \neg q$

which is the disjunction of the negated propositions. That is

$\neg(p \wedge q)$ **has the same meaning as** $\neg p \vee \neg q$

Notice that here we used the word 'both'. Like the word 'either', it is not a logical connective but it can be introduced with care to make the English more akin to everyday speech. Notice also the fact that the pencil is **not red or not sharp** does not exclude the possibility that it is **both not red and not sharp** – it is the inclusive **or** not the exclusive **or**.

Having seen that $\neg(p \wedge q)$ means the same as $\neg p \vee \neg q$ let us consider their truth tables:

p	q	$\neg$	(p	$\wedge$	q)	$(\neg p)$	$\vee$	$(\neg q)$
0	0	1	0	0	0	1	1	1
0	1	1	0	0	1	1	1	0
1	0	1	1	0	0	0	1	1
1	1	0	1	1	1	0	0	0

Notice that the truth values of each of these compound propositions are the same for the same truth values of their component propositions.

If two different compound propositions constructed from the same component propositions have the same truth values then they have the same meaning. Two compound propositions that have the same meaning are said to be **logically equivalent** to each other. As a further example consider the two compound propositions:

$\neg(p \vee q)$ and $(\neg p) \wedge (\neg q)$

These have the following truth tables:

p	q	$\neg$	(p	$\vee$	q)	$(\neg p)$	$\wedge$	$(\neg q)$
0	0	1	0	0	0	1	1	1
0	1	0	0	1	1	1	0	0
1	0	0	1	1	0	0	0	1
1	1	0	1	1	1	0	0	0

The list of truth values of the left-hand table is the same as the list of truth values of the right-hand table, that is both compound propositions have the same truth values. As a consequence they have the same meaning and are, therefore, logically equivalent. We write:

$\neg(p \wedge q) \equiv (\neg p) \vee (\neg q)$ **and** $\neg(p \vee q) \equiv (\neg p) \wedge (\neg q)$

where the symbol $\equiv$ stands for **logical equivalence**.

Conclusion: If two propositions have the same meaning they are said to be logically equivalent. To have the same meaning they must be composed of the

The symbol for equality (=) is often substituted for that of equivalence. While this is not strictly correct it is a commonly accepted substitution.

*same component propositions and have the same truth values for correspond-
ing truth values of their component propositions.*

Worked example 2.9

Prove the logical equivalence of each of the following compound propositions:

(a) $p \vee (q \wedge r)$ and $(p \vee q) \wedge (p \vee r)$

(b) $(\neg p) \vee (p \wedge q)$ and $(\neg p) \vee q$

(a)

p	q	r	p	∨	(q	∧	r)	(p	∨	q)	∧	(p	∨	r)
0	0	0	0	**0**	0	0	0	0	0	0	**0**	0	0	0
0	0	1	0	**0**	0	0	1	0	0	0	**0**	0	1	1
0	1	0	0	**0**	1	0	0	0	1	1	**0**	0	0	0
0	1	1	0	**1**	1	1	1	0	1	1	**1**	0	1	1
1	0	0	1	**1**	0	0	0	1	1	0	**1**	1	1	0
1	0	1	1	**1**	0	0	1	1	1	0	**1**	1	1	1
1	1	0	1	**1**	1	0	0	1	1	1	**1**	1	1	0
1	1	1	1	**1**	1	0	1	1	1	1	**1**	1	1	1

Because both compound propositions have identical truth listings we can
say that

$$p \vee (q \wedge r) \equiv (p \vee q) \wedge (p \vee r)$$

(b)

p	q	(¬p)	∨	(p	∧	q)	(¬p)	∨	q
0	0	1	**1**	0	0	0	1	**1**	0
0	1	1	**1**	0	0	1	1	**1**	1
1	0	0	**0**	1	0	0	0	**0**	0
1	1	0	**1**	1	1	1	0	**1**	1

Because the two propositions have identical truth listings we can say

$$(\neg p) \vee (p \wedge q) \equiv (\neg p) \vee q$$

Exercise

2.9 Prove the logical equivalence of each of the following compound
propositions:

(a) $p \vee (p \wedge q)$ and p

(b) $(p \wedge q) \vee [(r \vee p) \wedge \neg q]$ and $p \vee (r \wedge \neg q)$

The algebra of the three connectives
The three connectives AND, OR and NOT combine together according to a set of rules called their **algebra**. We shall merely state the rules here, their validity being left to you as an additional exercise. The rules are, for propositions p, q and r, as follows.

Commutativity
The order in which propositions are combined under conjunction or disjunction does not matter. For example

roses are red and violets are blue

means the same as

violets are blue and roses are red

That is

$$p \wedge q \equiv q \wedge p$$

Also

$$p \vee q \equiv q \vee p$$

We say that the logical connectives of conjunction ($\wedge$) and disjunction ($\vee$) are **commutative**.

Associativity
The order in which propositions are grouped does not matter. For example

I am old and I am grey and decrepit

means the same as

I am old and grey and I am decrepit

That is

$$(p \wedge q) \wedge r \equiv p \wedge (q \wedge r)$$

Also

$$(p \vee q) \vee r \equiv p \vee (q \vee r)$$

Conjunction and disjunction are **associative**.

Distributivity

This is similar to the distribution of multiplication over addition of the natural numbers. That is, multiplication takes precedence over addition and, by using brackets as we have done here, we can force the precedence of AND over OR. For example, the compound proposition

He is English and she is French or Belgian

is logically equivalent to

He is English and she is French or he is English and she is Belgian

That is

$$p \wedge (q \vee r) \equiv (p \wedge q) \vee (p \wedge r)$$

Also

$$p \vee (q \wedge r) \equiv (p \vee q) \wedge (p \vee r)$$

Identity

A proposition that is always true – can never be ascribed a value false – is called a **tautology** and is signified by the symbol t. A proposition that is always false – can never be ascribed a value true – is called a **contradiction** and is signified by the symbol c. Tautologies and contradictions form the following conjunctions and disjunctions:

$p \wedge t \equiv p$ **and** $p \wedge c \equiv c$

$p \vee t \equiv t$ **and** $p \vee c \equiv p$

The Latin word *idem* means the same.

Idempotency

The conjunction or disjunction of a proposition with itself is itself; no additional meaning is added:

$p \wedge p \equiv p$ **and** $p \vee p \equiv p$

Negation

The conjunction of a proposition with its negation is always false. The disjunction of a proposition with its negation is always true. For example

I shall go or I shall not go

is always true, whereas

I shall go and I shall not go

is always false. That is

$$p \wedge (\neg p) = c \text{ and } p \vee (\neg p) = t$$

DeMorgan's laws

Two important laws known as DeMorgan's laws we have already seen to be true.

$$\neg(p \wedge q) \equiv (\neg p) \vee (\neg q) \text{ and } \neg(p \vee q) \equiv (\neg p) \wedge (\neg q)$$

Involution

The negation of the negated proposition is the proposition. For example

The tree is not not hollow

is the same as

The tree is hollow

that is

$$\neg\neg p \equiv p$$

Absorption

$$p \vee (p \wedge q) \equiv p \text{ and } p \wedge (p \vee q) \equiv p$$

Here, the proposition q has been absorbed in each case. The validity of the absorption law is best demonstrated using equivalent switching circuits. In the switching circuit shown in Figure 2.1, current will flow across the circuit if switch A is ON and will not flow if switch A is OFF. The state of switch B is irrelevant. Consequently the parallelly connected circuit is equivalent to the single switch A, that is

$$A \vee (A \wedge B) \equiv A$$

or, in terms of propositions

$$p \vee (p \wedge q) \equiv p$$

In the switching circuit in Figure 2.2, current will flow across the circuit if switch A is ON and will not flow if switch A is OFF. The state of switch B is irrelevant. Consequently, the parallelly connected circuit is equivalent to the single switch A, that is

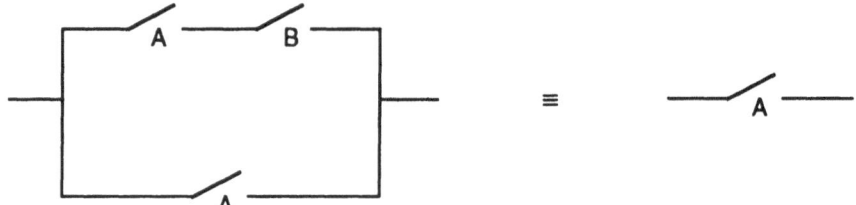

Figure 2.1

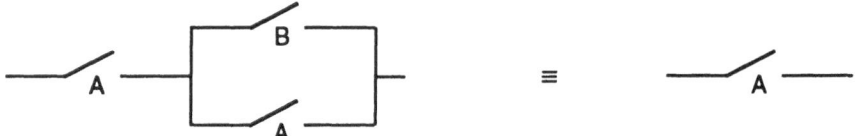

Figure 2.2

$$A \wedge (A \vee B) \equiv B$$

or, in terms of propositions

$$p \wedge (p \vee q) \equiv p$$

Worked example 2.10

Reduce the following compound propositions to simpler forms using the laws of logic:
(a) The applicant is at least 18 years old and a student or the applicant is at least 18 years old and unemployed.
(b) The customer is either given credit or he is a new customer and the customer is either given credit or he has a high-risk credit rating.

Solution:
(a) Let p = The applicant is at least 18 years old, q = The applicant is a student and r = The applicant is unemployed. The compound proposition can then be written as

$$(p \wedge q) \vee (p \wedge r) \equiv p \wedge (q \vee r)$$

by the **distributive law**. This reads as: The applicant is at least 18 years old and either a student or unemployed.
(b) Let p = The customer is given credit, q = The customer is a new customer and r = The customer has a high-risk credit rating. The

compound proposition can then be written as:

$(p \vee q) \wedge (p \vee r) \equiv p \vee (q \wedge r)$

by the **distributive law**. This reads as: The customer is either given credit or he is a new customer and he has a high-risk credit rating.

Exercise

2.10 Reduce the following compound propositions to simpler forms using the laws of logic:
    (a) Either the stock is above the reorder level or a warning is issued and either the stock is above the reorder level and a reorder is made.
    (b) I shall or shall not go but I shall find out for myself.

Truth tables and their simplification

Analysing truth tables
The compound proposition $(p \vee q) \wedge (p \vee r)$ has the following truth table:

p	q	r	(p	∨	q)	∧	(p	∨	r)	
0	0	0	0	0	0	**0**	0	0	0	
0	0	1	0	0	0	**0**	0	1	1	
0	1	0	0	1	1	**0**	0	0	0	
0	1	1	0	1	1	**1**	0	1	1	$[q \wedge r]$
1	0	0	1	1	0	**1**	1	1	0	$[p]$
1	0	1	1	1	0	**1**	1	1	1	$[p \wedge r]$
1	1	0	1	1	1	**1**	1	1	0	$[p \wedge q]$
1	1	1	1	1	1	**1**	1	1	1	$[p \wedge q \wedge r]$

Alongside the table are listed a number of component propositions. If any one of these component propositions is true then the complete proposition is also true. However, the converse does not follow. The truth of the complete proposition does not guarantee the truth of a particular component proposition. For example, the complete proposition is true if the component proposition p is true and the component proposition $q \wedge r$ is false. What can be stated is that if the complete proposition is true then one of these component propositions is also going to be true. That is

$(q \wedge r) \vee p \vee (p \wedge r) \vee (p \wedge q) \vee (p \wedge q \wedge r)$

is true when

$(p \lor q) \land (p \lor r)$

is true and vice versa, and

$(q \land r) \lor p \lor (p \land r) \lor (p \land q) \lor (p \land q \land r)$

is false when

$(p \lor q) \land (p \lor r)$

is false and vice versa.

As a consequence the two compound propositions are logically equivalent – they have identical truth values for corresponding truth values of their component propositions:

$(q \land r) \lor p \lor (p \land r) \lor (p \land q) \lor (p \land q \land r) \equiv (p \lor q) \land (p \lor r)$

Notice that the proposition on the left is a conjunction of component propositions. Now, by using the laws of logic, we can reduce the proposition on the left to a simpler form:

$p \lor (p \land r) = p$

by the absorption law. Hence the left-hand proposition can be written as

$(q \land r) \lor p \lor (p \land q) \lor (p \land q \land r)$

Again, because $p \lor (p \land q) \equiv p$ and $p \lor (p \land q \land r) \equiv p$ by the absorption law the truth function can be written as

$(q \land r) \lor p \equiv p \lor (q \land r)$

using the commutative law. Hence we have reduced $(p \lor q) \land (p \lor r)$ to the equivalent and simpler form $p \lor (q \land r)$. This will be recognized as the distributive law that we met previously.

Demorganization

Every compound proposition has an associated, negated compound proposition that is found by a process of **negation** known as **demorganizing**. When a compound proposition is demorganized every component of the compound proposition is negated (with double negations being cleared), conjunctions are changed to disjunctions and disjunctions are changed to conjunctions. For example, the demorganization of the expression

$p \land (q \lor r)$

yields

$$\neg p \vee (\neg q \wedge \neg r)$$

and, by the distributive law that we have just demonstrated, this compound proposition can be written as

$$(\neg p \vee \neg q) \wedge (\neg p \vee \neg r)$$

If this expression is demorganized we find that it yields

$$(p \wedge q) \vee (p \wedge r)$$

Demorganizing twice has the effect of negating the original compound proposition twice, so reverting it to its original sense. In other words, we have shown that

$$(p \wedge q) \vee (p \wedge r) \equiv p \wedge (q \vee r)$$

thus demonstrating the second distributive law.

Simplifying compound propositions

When dealing with propositions and their connectives it is often found that very large and unwieldy compound propositions are built up. We shall see here how such compound propositions can be reduced to simpler, more manageable, forms.

The basic principle is to convert the compound proposition being dealt with into a conjunction of terms and then to use the laws of logic to reduce it to a simpler form. For example, the compound proposition (already a conjunction of terms)

$$((p \wedge \neg q) \vee (\neg p \wedge r)) \vee (p \wedge q)$$

can be written as

$$((\neg p \wedge r) \vee (p \wedge \neg q)) \vee (p \wedge q)$$

using the commutativity law. This becomes

$$(\neg p \wedge r) \vee ((p \wedge \neg q) \vee (p \wedge q))$$

by the associativity law and

$$(\neg p \wedge r) \vee (p \wedge (\neg q \vee q))$$

by the distributivity law and

> Do not be surprised if you find this aspect difficult. Using the algebra of propositions to simplify expressions requires a deal of experience which can only be gained by working through the examples.

$(\neg p \wedge r) \vee (p \wedge t)$ because $\neg q \vee q \equiv t$

by the negation law and

$(\neg p \wedge r) \vee p$

by the identity law and

$(\neg p \vee p) \wedge (r \vee p)$

by the distributive law and

$t \wedge (r \vee p)$

by the negation law and

$(r \vee p)$

by the identity law and

$(p \vee r)$

by the commutativity law. Consequently,

$$[(p \wedge \neg q) \vee (\neg p \wedge r)] \vee (p \wedge q) \equiv (p \vee r)$$

The original expression on the left can be replaced by the equivalent, simpler form on the right.

Worked example 2.11

If, at a hole-in-the-wall machine, your bank card number is not verified or you enter your personal identification number (PIN) incorrectly then you will not be granted access to your account. Let CardOk be a Boolean variable that is TRUE if your card is verified and let PinOk be a Boolean variable that is TRUE if you entered your PIN correctly. Is the following program segment sufficient to safeguard your account?

```
If (CardOk AND NOT PinOk) or (NOT CardOk AND PinOk)
        Then NoAccess
        Else Access
```

where Access is a procedure to access your account and NoAccess is a procedure to terminate your enquiry.

Solution: Let p = CardOk and q = PinOk. The conditional statement in the implication can then be written in symbolic form as

$$(p \wedge \neg q) \vee (\neg p \wedge q)$$

The truth table for this compound proposition is

p	q	(p	∧	¬q)	∨	(¬p	∧	q)
0	0	0	0	1	**0**	1	0	0
0	1	0	0	0	**1**	1	1	1
1	0	1	1	1	**1**	0	0	0
1	1	1	0	0	**0**	0	0	1

From the truth table we can see that if the compound proposition is TRUE (1) then NoAccess is permitted, but if the proposition is FALSE (0) then Access is permitted. Consequently, access to your account can occur if both Boolean variables are FALSE (p = 0 and q = 0).

This means that access to your account is permitted if both the card number fails to be verified and an incorrect PIN is entered. This unsatisfactory state of affairs can be rectified by gleaning the condition r as follows:

p	q	r	¬r
0	0	1	0
0	1	1	0
1	0	1	0
1	1	0	1

Here r is FALSE only when both p and q are TRUE – when r is FALSE then Access is permitted. From the list of truth values we can see that:

$$\neg r \equiv (p \wedge q)$$

which means that

$$r \equiv \neg(p \wedge q) \equiv (\neg p \vee \neg q)$$

```
If (NOT CardOk OR NOT PinOk)
        Then NoAccess
        Else Access
```

Worked example 2.12

A company credit manager has issued instructions to the sales personnel that govern the referral of a sale to the credit office prior to despatch.

A sale must be referred to the credit office before it is despatched if:

(a) The customer has a good credit rating but has an outstanding balance due (regardless of whether or not the new order is within the limit).

or

(b) The customer has no outstanding balance but places an order outside the limit (regardless of whether or not the customer has a good credit rating).

Is this a good set of instructions that could be programmed or can you suggest better?

Solution: Let p = customer credit rating good, q = customer has an outstanding balance due and r = customer has placed an order within the limit. The conditions can then be written in symbolic form as

(a) $(p \wedge q \wedge r) \vee (p \wedge q \wedge \neg r) \equiv (p \wedge q)$ by the absorption law
(b) $(p \wedge \neg q \wedge \neg r) \vee (\neg p \wedge \neg q \wedge \neg r) \equiv (\neg q \wedge \neg r)$ by the absorption law

Consequently, a sale must be referred to the credit office if either

$(p \wedge q)$ is TRUE

or

$(\neg q \wedge \neg r)$ is TRUE

Combining these conditions we can say that if the compound proposition:

$(p \wedge q) \vee (\neg q \wedge \neg r)$

is TRUE (1) – that is the conditions are met – then a sale must be referred to the credit office prior to its despatch. If this compound proposition is FALSE (0) – that is the conditions are not met – then the sale need not be referred to the credit office prior to its despatch. The truth table for this compound proposition is given as

p	q	r	(p	∧	q)	∨	(¬q	∧	¬r)	
0	0	0	0	0	0	1	1	1	1	
0	0	1	0	0	0	0	1	0	0	*
0	1	0	0	0	1	0	0	0	1	*
0	1	1	0	0	1	0	0	0	0	*
1	0	0	1	0	0	1	1	1	1	
1	0	1	1	0	0	0	1	0	0	
1	1	0	1	1	1	1	0	0	1	
1	1	1	1	1	1	1	0	0	0	

From lines 2 to 4 of this table (those marked by a *) we can see that there are three cases where no reference is made to the credit office despite the fact that the customer does not have a good credit rating (p = 0). Consequently, the instructions are flawed because goods can be despatched to a customer without a credit rating with no reference to the credit office. To prevent this situation occurring a better set of conditions must be devised.

Let the improved set of conditions be embodied within a compound proposition s where s and $\neg s$ have the following truth table:

p	q	r	s	$\neg s$	
0	0	0	1	0	
0	0	1	1	0	*
0	1	0	1	0	*
0	1	1	1	0	*
1	0	0	1	0	
1	0	1	0	1	
1	1	0	1	0	
1	1	1	1	0	

Note that reference must be made to the credit office when s is TRUE.

Here we have retained the earlier conditions (when s = 1) but have also imposed the additional conditions that reference must be made to the credit office when the customer does not have a good credit rating, that is s = 1 now when p = 0.

From this table we can see that $\neg s$ is only true for one case, that is

$$\neg s \equiv p \wedge \neg q \wedge r$$

Hence

$$s \equiv \neg(p \wedge \neg q \wedge r) \equiv (\neg p \vee q \vee \neg r)$$

by demorganization.

Because reference must be made to the credit office when s is TRUE, the improved set of rules is seen to be:

A sale is referred to the credit office if the customer:

(a) has no good credit rating

or

(b) has an outstanding balance due

or

(c) places a new order outside the limit.

Worked example 2.13

In a bid to encourage Sunday playing a local golf club arranged a tournament for the second Sunday in May. The club committee decided that the participants should only be selected from
(a) those club members who did not play on Sundays whether or not they played on Saturday;
(b) those club members who played on both Saturdays and Sundays.
The latter is to encourage the former. Can you suggest a simpler rule?

Solution: Let p = club member, q = Sunday player and r = Saturday player. The conditions for selection to play in the tournament can then be written in symbolic form as
(a) $(p \wedge \neg q \wedge r) \vee (p \wedge \neg q \wedge \neg r)$
(b) $(p \wedge q \wedge r)$
The complete set of conditions can then be written as

$(p \wedge \neg q \wedge r) \vee (p \wedge \neg q \wedge \neg r) \vee (p \wedge q \wedge r)$

$\equiv (p \wedge \neg q) \vee (p \wedge q \wedge r)$	by the absorption law
$\equiv [(p \wedge \neg q) \vee (p \wedge q)] \wedge [(p \wedge \neg q) \vee r]$	by the distributive law
$\equiv p \wedge [(p \wedge \neg q) \vee r]$	by the absorption law
$\equiv [p \wedge (p \wedge \neg q)] \vee [p \wedge r]$	by the distributive law
$\equiv (p \wedge \neg q) \vee (p \wedge r)$	by the idempotency law
$\equiv p \wedge (\neg q \vee r)$	by the distributive law

A club member who either does not play on Sunday or who does play on Saturday.

Exercises

2.11 Entry to a computer network is gained by a user entering a user name followed by a password. Let User be a Boolean variable that is TRUE if a valid user name is entered and Pass be a Boolean variable that is TRUE if a correct password is entered. Does the following program segment properly control access to the network?

```
If NOT (User OR Pass)
        Then NoAccess
        Else Access
```

2.12 The following rules apply to the assessment of a course in a certain college:

A pass will be awarded if the candidate:

(a) obtains 50% or over in the examination and 50% or over in the in-course assessment;

(b) obtains up to 50% in the examination and is exempted from the in-course assessment;

(c) obtains 50% or over in the examination and up to 50% in the in-course assessment.

Can you simplify the rules?

2.13 A garage placed the following advertisement in the local paper:

Apprentice mechanic required: must be at least 21 years old and without experience or under 21 years with experience. Will also consider those 21 years or older with experience and those under 21 years without experience.

Who can apply?

___

Converse, inverse and contrapositive

Converse
Given the implication p → q the implication q → p is referred to as the **converse** of the original form. For example, the converse of the implication:

If I discuss pigs then I talk about animals

is

If I talk about animals then I discuss pigs

Notice that the truth or falsity of the converse does not necessarily follow from the truth or falsity of the original implication. The set of pigs forms a subset of the set of animals, so to talk about animals does not mean that you are necessarily talking about pigs. This is easily seen to be true by demonstrating that:

$$(p \to q) \to (q \to p)$$

is not a tautology.

Inverse
The implication ¬p → ¬q is called the **inverse** of the implication p → q. For example, the inverse of the implication:

If I discuss pigs then I talk about animals

is

If I do not discuss pigs then I do not talk about animals

Notice again that the truth or falsity of the inverse does not necessarily follow from the truth or falsity of the original implication. By not discussing pigs does not necessarily mean that you are not talking about animals. This is easily seen to be true by demonstrating that:

$$(p \rightarrow q) \rightarrow (\neg p \rightarrow \neg q)$$

is not a tautology.

Contrapositive

The implication $\neg q \rightarrow \neg p$ is called the **contrapositive** of the implication $p \rightarrow q$. For example, the contrapositive of the implication:

If I discuss pigs then I talk about animals

is

If I do not talk about animals then I do not discuss pigs

Here, the truth or falsity of the contrapositive does not necessarily follow from the truth or falsity of the original implication. This is easily seen to be true by demonstrating that:

$$(p \rightarrow q) \rightarrow (\neg q \rightarrow \neg p)$$

is a tautology.

Worked example 2.14

Demonstrate the equivalence between an implication and its contrapositive:

$$\neg q \rightarrow \neg p \equiv p \rightarrow q$$

Solution:

p	**q**	**¬q**	**→**	**¬p**	**p**	**→**	**q**
0	0	1	**1**	1	0	**1**	0
0	1	0	**1**	1	0	**1**	1
1	0	1	**0**	0	1	**0**	0
1	1	0	**1**	0	1	**1**	1

The identical truth tables demonstrate the logical equivalence of the two implications.

Exercise

2.14 Show that the inverse of an implication is the contrapositive of the converse of the implication and that the inverse is logically equivalent to the converse.

Part Two

Sets, Lists and Counting

We all know what a whole number is and we all know how to add them together. This is just as well, because if we were to sit down and try to think of how to define a whole number and how to define addition we would be sitting down for a long time. The concept of a whole number and the concept of addition of two whole numbers are notions that are taken as being commonly understood and capable of being accepted without definition. Any definition of a concept rests upon earlier definitions, and in the case of whole numbers and addition there are no earlier concepts.

Once we accept whole numbers and addition as being intuitively understood we can then proceed to define subtraction as the reverse process of addition, multiplication as repetitive addition and division as the reverse process of multiplication – repetitive subtraction. Raising to a power is defined as repetitive multiplication, and with a few interspersed definitions and adjustments it becomes possible to derive the concepts of rational and irrational numbers – the totality of the real number system and its arithmetic. The entire panoply of arithmetic, the very foundation of mathematics, rests upon two undefined ideas.

We cannot communicate this kind of structured knowledge to a computer because a computer does not work in the same way that we do. We can, of course, make a computer perform many of the functions that we perform but not necessarily in the same way that we would naturally perform them. We often have to reorientate our way of thinking about things so that a computer can then accommodate the processes and thereby emulate our abilities.

In Part Two we exemplify the principle of reorientating our thinking about whole numbers and the various processes involved in using whole numbers to count.

Chapter 3

Sets and numbers

OBJECTIVES

When you have completed this chapter you will be able to:

- ☐ distinguish between data and information;

- ☐ define data types using the language of sets;

- ☐ construct subsets and the power set of any other set;

- ☐ create the union and intersection of two or more sets;

- ☐ define an appropriate universal set and find the complement of a set;

- ☐ partition a set into a disjoint union;

- ☐ display the union and intersection of two or more sets on a Venn diagram;

- ☐ use a Venn diagram to prove statements concerning sets;

- ☐ use a Venn-type diagram to illustrate truth and falsity in the propositional calculus;

- ⊓ construct the data type of natural numbers from the atomic data type of **sets**;

- ☐ define addition and subtraction using the operations of successor and predecessor.

So far we have seen how our ideas concerning the processes of reasoning can be put on a more formal basis by using symbols. This not only has the advantage of simplifying our descriptions by clearing away unwanted detail, but it also displays the distinct advantage of considering general situations rather than specific instances. Indeed, the notion of symbol is fundamental to mathematical thinking where it is the pattern, the regularity, the commonality that is looked for rather than the specific.

The mathematician's desire to classify and generalize is complemented by the desire to ensure that all our descriptions be free of ambiguity and understood by all. This can only be achieved by a commonly understood language: the language of mathematics. At the very heart of both the language and the desire to draw together commonalities is the notion of a set. Consequently, we must be familiar with the language of set theory and the manipulative use of sets for our later work.

Data types and sets

Computers thrive on data. Indeed, computers are pretty useless without it. But what exactly is data?

Anyone who uses a computer is interacting with it. From typing in instructions at the C prompt in DOS to using a mouse to point to a specific area of a screen; every time a user causes the computer to change from the state that it is in to another state, data has entered the computer. Often the word 'data' is confused with the word 'information', and many times the two are taken to mean the same thing. There is, however, a difference between data and information.

Data
A datum (for which the plural is data) consists of a sequence of individual **characters**. A character can exist on a keyboard. For example, if the computer is displaying the DOS prompt and you depress the letter A on the keyboard then the letter A will appear on the monitor screen – you have entered a datum into the computer. A character may not exist explicitly on a keyboard – it may require a combination of keys to be pressed. Typical of such characters are graphic characters, such as lines that require the use of an alphabetic key with the Control key or the Alt key.

Information
Information is data accompanied with a context **or** data with a meaning that the computer understands. For example, if, at the DOS prompt, you entered the sequence of characters

    dir c:\

followed by pressing the Enter key the computer would then read the contents listing of the root directory on the hard disk C and display it on the monitor screen. Here the data – the string of characters that you typed in at the keyboard – was recognized by the computer as an instruction: it had meaning that the computer understood. This is information.

There are a number of definitions of information, and while the above definition of information is somewhat arcane it will suffice for our purposes as we shall not be dealing with it in any greater detail within the pages of this book. What is important is that we have distinguished between information and data.

Conclusion: A datum consists of a sequence of characters. Information is data coupled with a context.

Data types

We have all heard stories about someone who has received a computerized bill for £0.00 and ignored it only to have it followed up with a writ for non-payment. Whether such stories are true or are the figment of a fevered imagination is beside the point. What is relevant is that the story highlights a crucial problem that exists among the various activities relating to the processing of data. Data must be verified. That is, if the computer program is prompting for numeric data then only numeric data must be provided; if the computer is prompting for one of the first three upper-case letters of the alphabet, only one of A, B or C must be provided. However, if the data is provided by, for example, a human operator then there is always a possibility that incorrect data could be entered. This is where verification comes in. Any data entered must be checked that it is within the prescribed bounds and that it is of the correct type.

The ubiquitous language Basic makes a first attempt at this by distinguishing between numeric variables and string variables. While any keyboard character can be entered into a string variable, only numbers can be entered into a numeric variable. More sophisticated languages such as Pascal, Modula 2 and C use the concept of a data type that can be defined within the program itself. For example, the following Pascal code placed with a program will define a data type called Seasons:

```
type        Seasons = (Spring, Summer, Autumn, Winter);
var         TimeOfYear: Seasons;
```

The variable TimeOfYear is now a variable of **type** Seasons and if, at any time after the declaration of this data type, a user is required to enter a value into the variable TimeOfYear then the program will permit only one of the four listed values to be entered. When the variable is assigned one of the data type values it is said to have been **instantiated** and there are as many different possible instantiations of the variable as there are data type values.

This description of a data type is rather rudimentary. Strictly speaking, a data type needs to be accompanied by a collection of operations that can be used to manipulate the instantiated values. However, for our current purposes it is sufficient to restrict our description to what has been said.

Data types are fundamental to programming and the ability to use and manipulate data types requires a knowledge of the mathematical theory of sets.

Conclusion: A declaration within a program that associates data of a particular type to a declared name is called a data type.

Sets

A **set** is a collection of objects such as a set of tools or a set of numbers or a set of values of a data type. The notation used to denote a set consists of a pair of curly brackets {...} with the individual objects that constitute the set described inside the brackets. For example, the set consisting of the whole numbers 1 to 6 can be written as:

$$\{1, 2, 3, 4, 5, 6\}$$

where the contents are **listed** individually. Alternatively, the set could be written as

$$\{x: x \text{ is a whole number and } 1 \leq x \leq 6\}$$

This is read as:

The set of x values, where the value of x is a whole number and x is greater than or equal to 1 and less than or equal to 6.

The colon (:) stands for the word **where**. In this description of the set the contents are **prescribed** by describing the properties that they hold in common. This description of the set contents is also referred to as the **set builder** description because the specific contents of the set can be built up from the prescription. Notice that the order in which the elements are listed does not matter, nor does the repetition of elements as each element is taken as unique. For example, the sets

$$\{p, q, r\}, \{q, r, p\}$$

and

$$\{r, r, r, q, q, p, q\}$$

are all the same set.

Many times it is more convenient to refer to a set without specifically describing its contents, in which case we use an **identifier**. For example

$$A = \{\text{head, tail}\}$$

Here the set consisting of the possible outcomes from tossing a coin has been called set A and can now be referred to in future by simply using the letter A as identifier. Note that more extended identifiers are generally used within a computer program. For example

$$\text{CoinToss} = \{\text{head, tail}\}$$

This is done for clarity in the reading of the program. However, within what follows we shall restrict our identifiers to be shorter as an aid to brevity.

The individual objects that are contained within a set are called **elements** of the set, and the symbol $\in$ is used to denote membership of a set. For example, if

A = {1, 2, 3}

and

B = {3, 4, 5}

then $1 \in$ A and $5 \in$ B. That is, 1 **is an element of** set A and 5 **is an element of** set B. Notice that element $1 \notin$ B and $5 \notin$ A: the slash through the symbol negates the symbol so that $\notin$ means **not a member**. Hence, 1 **is not an element of** B and 5 **is not an element of** A.

Set notation is used to describe data types. For example, in C the data type **seasons** is defined, or enumerated, as follows:

```
enum seasons {Spring, Summer, Autumn, Winter};
```

Notice the use in C of the curly brackets { .. }, whereas in Pascal round brackets (..) are used.

Subsets
Additional sets can be formed by using the elements of a given set. For example, given the set

A = {1, 2, 3}

then by just using the elements of set A the following additional sets could be formed:

{1}, {2}, {3}, {1, 2}, {1, 3}, {2, 3}

Each of these sets is called a **subset** of A, and the notation used to denote a subset is $\subseteq$. For example

{1, 2} $\subseteq$ {1, 2, 3}

the set {1, 2} is a subset of the set {1, 2, 3}.

Every set also has the **empty set** as a subset. The empty set is the set that contains no elements at all and is denoted by

{}

or, alternatively, by

$$\varnothing$$

The empty set may seem to be an odd set to define but, as will be demon-strated later, it is necessary to complete the rules which govern the manipu-lation of sets – just as it is necessary to define the number 0 to enable num-bers to be manipulated.

Conclusion: A set is a collection of elements and a subset of a set is a col-lection of elements selected from the set.

Proper subsets
In addition to the fact that every set has the empty set as a subset, every set also has itself as a subset. That is

$$\{1, 2, 3\} \subseteq \{1, 2, 3\}$$

If a subset contains fewer elements than the set of which it is a subset then it is called a **proper** subset and the symbol $\subset$ is used to denote this. For exam-ple

$$\{1, 2\} \subset \{1, 2, 3\}$$

hence the underline beneath the subset symbol $\subseteq$, which can be taken to mean $\subset$ **or** =. For example

$$\{1, 2\} \subseteq \{1, 2, 3\}$$

can be read as

$$\{1, 2\} \subset \{1, 2, 3\} \text{ or } \{1, 2\} = \{1, 2, 3\}$$

Notice that here are two propositions connected by OR, making the com-pound proposition TRUE if either component proposition is TRUE.

Power sets
The **power set** P(A) of set A is the set whose elements are all the subsets of A. For example, the power set of the set $\{H, T\}$ is the set

$$P(\{H, T\}) = \{\varnothing, \{H\}, \{T\}, \{H, T\}\}$$

Notice that the power set contains both the original set and the empty set.

Set cardinality

The cardinality of a set is the number of elements that it contains. Of particular interest is the cardinality of a power set P(A). For example, the power set of the set {1} is

$$P(\{1\}) = \{\varnothing, \{1\}\}$$

which has cardinality 2. The power set of the set {1, 2} is

$$P(\{1, 2\}) = \{\varnothing, \{1\}, \{2\}, \{1, 2\}\}$$

which has cardinality 4 or 2^2, where the power 2 is the cardinality of {1, 2} and the power set of the set {1, 2, 3} is

$$P(\{1, 2, 3\}) = \{\varnothing, \{1\}, \{2\}, \{3\}, \{1, 2\}, \{1, 3\}, \{2, 3\}, \{1, 2, 3\}\}$$

which has cardinality 8 or 2^3, where the power 3 is the cardinality of {1, 2, 3}. In general, the power set of set A has cardinality 2^n, where n is the cardinality of set A.

Conclusion: A proper subset of a set is wholly contained within the set. The power set is the set of all possible subsets of the set and the set cardinality is the number of elements within the set.

While the argument that we have used to define the cardinality of a power set seems reasonable it does not constitute a proof. A proof will be given later in the book.

Worked example 3.1

Write down, in symbolic form, the following sets:
(a) the set of whole numbers between 5 and 9 inclusive;
(b) the set of numbers strictly between 5 and 9;
(c) the set of lower-case vowels.

Solution:
(a) {5, 6, 7, 8, 9}
(b) $\{x: x \text{ is a number and } 5 < x < 9\}$
(c) {a, e, i, o, u}

Worked example 3.2

Find
 (i) all proper subsets of the set A;
(ii) the power set of set A and the cardinality of the power set.
where
(a) A = {a, b, c}

(b) A = {x: x is a vowel}

(c) A = {red, white, blue}

Solution:

(ia) ∅, {a}, {b}, {c}, {a, b}, {a, c}, {b, c}

(iia) P({a, b, c}) = {∅, {a}, {b}, {c}, {a, b}, {a, c}, {b, c}, {a, b, c}}, cardinality = $2^3 = 8$

(ib) ∅, {a}, {e}, {i}, {o}, {u}, {a, e}, {a, i}, {a, o}, {a, u}, {e, i}, {e, o}, {e, u}, {i, o}, {i, u}, {o, u}, {a, e, i}, {a, e, o}, {a, e, u}, {a, i, o}, {a, i, u}, {a, o, u}, {e, i, o}, {e, i, u}, {e, o, u}, {i, o, u}, {a, e, i, o}, {a, e, i, u}, {a, e, o, u}, {a, i, o, u}, {e, i, o, u}

(iib) P({a, e, i, o, u}) = {∅, {a}, {e}, {i}, {o}, {u}, {a, e}, {a, i}, {a, o}, {a, u}, {e, i}, {e, o}, {e, u}, {i, o}, {i, u}, {o, u}, {a, e, i}, {a, e, o}, {a, e, u}, {a, i, o}, {a, i, u}, {a, o, u}, {e, i, o}, {e, i, u}, {e, o, u}, {i, o, u}, {a, e, i, o}, {a, e, i, u}, {a, e, o, u}, {a, i, o, u}, {e, i, o, u}, {a, e, i, o, u}}, cardinality = $2^5 = 32$

(ic) ∅, {red}, {white}, {blue}, {red, white}, {red, blue}, {white, blue}

(iic) P({red, white, blue}) = {∅, {red}, {white}, {blue}, {red, white}, {red, blue}, {white, blue}, {red, white, blue}}, cardinality = 8

Worked example 3.3

X is declared to be a variable of data type D. How many possible instantiations of *X* are there when?

(a) D = {1, 2, 3, 4}

(b) D = {x: x is a vowel}

(c) D = {x: x is a real number}

Solution:

(a) 4

(b) 5

(c) an infinite or uncountable number

Worked example 3.4

What is the power set of { }?

Solution: The empty set {} or ∅ is a subset of every set – including the empty set. Now you see why a set has to be a subset of itself. The power set of set A is the set of all subsets of set A, hence

$$P(\{ \}) = \{\{ \}\}$$

the set whose only element is the empty set. Notice that here the empty set is both an element and a subset of {{ }}.

Exercises

3.1 Write down in symbolic form, the following sets:
    (a) the set of lower-case letters between t and z;
    (b) the set of dates between 1 January and 31 January;
    (c) the set of days of the week.

3.2 Find
    (i) all proper subsets of the set A;
    (ii) the power set of set A and the cardinality of the power set;
    where
    (a) A = {10, 100, 1000}
    (b) A = {x: x is a suit of a deck of playing cards}
    (c) A = {Brie, Camembert, double Gloucester, Cheddar, Wensleydale}

3.3 X is declared to be a variable of data type D. How many possible instantiations of X are there when?
    (a) D = {d: d is an even integer and $-6 \leq d < 4$}
    (b) D = {d: d is a consonant}
    (c) D = {x: x is a rational number}

3.4 What is the power set of {{ }}?

Set operations

The union of two sets
The elements of two sets can be combined together to form a third set. For example, the elements of sets

   A = {Andrew, Colin, Eric, Thomas}

and

   B = {Eric, Thomas, William}

are boys' names and they can be combined together to form the set C, where

   C = {Andrew, Colin, Eric, Thomas, William}

Notice that the elements **Eric** and **Thomas** which are common to both sets have not been repeated in accordance with standard set notation.

This process of combining or **uniting** the elements of two sets is referred to as the **union** of two sets and is represented by the symbol $\cup$. We write

{Andrew, Colin, Eric, Thomas} $\cup$ {Eric, Thomas, William} = {Andrew, Colin, Eric, Thomas, William}

or, symbolically

$A \cup B = C$

It does not matter if set A is unioned with set B or set B is unioned with set A, the result is the same:

$A \cup B = B \cup A$

That is, the operation of union is **commutative**. The operation of union is also **associative**:

Here we see the beginnings of an algebra. We have our objects – **sets** – an operation between objects – **union** – and two rules that the objects and the operation obey.

$(A \cup B) \cup C = A \cup (B \cup C) = A \cup B \cup A$

Conclusion: The union of two sets is the set of elements that are in either set.

Worked example 3.5

Find the union C of sets A and B where
(a) A = {rose, daffodil, tulip, begonia}, B = {rose, begonia, fuchsia, poppy}
(b) A = {1, 3, 5, 7, 9}, B = {2.2, 4.5, 6, 8.3, 9}
(c) A = {d: d is a date and 1 March $\leq d < $ 1 June}, B = {s: s is a date and 1 May $< s \leq$ 14 August}

Solution:
(a) $A \cup B$ = {rose, daffodil, tulip, begonia, fuchsia, poppy}
(b) $A \cup B$ = {1, 2.2, 3, 4.5, 5, 6, 7, 8.3, 9}
(c) $A \cup B$ = {t: t is a date and 1 March $\leq t \leq$ 14 August}
Notice here that the use of d as a variable in A and the use of s as a variable in B is of no account. It is not the variable label that matters but the values, namely the dates themselves.

Worked example 3.6

Show that the union of all the subsets of:

A = {a, aa}

is A.

Solution: There are $2^2 = 4$ subsets of {a, aa} and they are:

∅, {a}, {aa}, {a, aa}

Hence

∅ ∪ {a} ∪ {aa} ∪ {a, aa} = [∅ ∪ {a}] ∪ [{aa} ∪ {a, aa}] = [{a}] ∪ [{a, aa}] = {a, aa}

Note that for any set A

A ∪ ∅ = A

Exercises

3.5 Find the union C of sets A and B where:
   (a) A = {envelope, paper, pen, pencil}, B = {paper, paper clip, stapler, pen}
   (b) A = {x: x is a rational number}, B = {y: y is an irrational number}
   (c) A = {t: t is a time and noon ≤ t < 4.00 p.m.}, B = {s: s is a time and 2.15 p.m. < s ≤ 6.30 p.m.}

3.6 Show that the union of all the subsets of:

A = {01, 10, 11}

is A.

The intersection of two sets
A third set can also be created from those elements that are common to two sets. For example, the two sets

A = {Andrew, Colin, Eric, Thomas} and B = {Eric, Thomas, William}

have the elements Eric and Thomas in common and these two elements can be used to define the set C:

C = {Eric, Thomas}

This process of selecting the common elements of two sets is referred to as the process of **intersection** and is represented by the symbol ∩:

{Andrew, Colin, Eric, Thomas} ∩ {Eric, Thomas, William} = {Eric, Thomas}

or, symbolically

A ∩ B = C

It does not matter if set A is intersected with set B or set B is intersected with set A, the result is the same:

A ∩ B = B ∩ A

That is, the operation of intersection is commutative. The operation of intersection is also associative:

(A ∩ B) ∩ C = A ∩ (B ∩ C) = A ∩ B ∩ A

Conclusion: The intersection of two sets is the set that consists of those elements common to both sets.

___

Worked example 3.7

Find the intersection C of sets A and B where
(a) A = {rose, daffodil, tulip, begonia}, B = {rose, begonia, fuchsia, poppy}
(b) A = {1, 3, 5, 7, 9}, B = {2.2, 4.5, 6, 8.3, 9}
(c) A = {d: d is a date and 1 March $\leq d <$ 1 June}, B = {s: s is a date and 1 May $< s \leq$ 14 August}

Solution:
(a) A ∩ B = {rose, begonia}
(b) A ∩ B = {9}
(c) A ∩ B = {t: t is a date and 1 May $< t \leq$ 31 May}. Notice that 1 May $< t$ is the same as 2 May $\leq t$

Worked example 3.8

Show that the intersection of all the subsets of:

A = {a, aa}

is ∅.

Solution: There are $2^2 = 4$ subsets of {a, aa} and they are:

∅, {a}, {aa}, {a, aa}

Hence

$$∅ \cap \{a\} \cap \{aa\} \cap \{a, aa\} = (∅ \cap \{a\}) \cap (\{aa\} \cap \{a, aa\}) = ∅ \cap \{aa\}$$
$$= ∅$$

Note that for any set A

$$A \cap ∅ = ∅$$

Exercises

3.7 Find the intersection C of sets A and B where
- (a) A = {envelope, paper, pen, pencil}, B = {paper, paper clip, stapler, pen}
- (b) A = {x: x is a rational number}, B = {y: y is an irrational number}
- (c) A = {t: t is a time and noon $\leq t <$ 4.00 p.m.}, B = {s: s is a time and 2.15 p.m. $< s \leq$ 6.30 p.m.}

3.8 Show that the intersection of all the subsets of

$$A = \{01, 10, 11\}$$

is ∅.

The universal set

When manipulating sets it is often useful to know not just which elements are in set A, but also those elements that are **not** in set A. This, however, can present a problem. For example, if set A contains the first five whole numbers:

$$A = \{1, 2, 3, 4, 5\}$$

then sct A docs not contain the numbers 6, −8, 2/3, π or even red apples. While it is useful to know that A does not contain 6, −8, 2/3 or π it is something of a nonsense to say that it does not contain red apples. After all, why red apples? Well, why not? Just because we have said that set A contains the first five whole numbers does not preclude the fact that it does not contain red apples or even red herrings for that matter. To exclude the possibility of nonsenses such as this one we must put some sort of bound to our discussions about set A; we must define the **type** of elements that we are talking about and this we do by defining a **universal set**.

In any discussion involving sets the universal set is the set that contains all those elements of the type under discussion. For example, if

$$A = \{1, 2, 3, 4, 5\}$$

then the universal set, denoted by S, could be the set of natural numbers.

Notice that the universal set is not unique. We could equally have chosen it to be the set of rational numbers or even the set of real numbers; the choice of S depends upon the context in which set A is being discussed.

The complement of a set depends upon the defined universal set and, accordingly, is unique only for a given universal set.

The complement of a set

The complement of set A is the set that contains all those elements in the universal set that are not in A. This set is denoted by A'. For example, if

$$S = \{0, 1, 2, 3, 4, 5, 6, 7, 8, 9\} \text{ (the decimal numerals)}$$

and

$$A = \{0, 2, 4, 6, 8\}$$

then

$$A' = \{1, 3, 5, 7, 9\}$$

Notice that

$$A \cup A' = S$$

that is the union of any set with its complement is the universal set and

$$A \cap A' = \varnothing$$

the intersection of any set with its complement is the empty set.

Conclusion: The universal set consists of all elements of the type under discussion, and any set under discussion is a subset of the universal set. The complement of a set is a set that consists of all those elements of the universal set that are not elements of the set.

Worked example 3.9

Define suitable universal sets for each of the following:
(a) $A = \{0.1, 0.01, 0.001, 0.0001\}$ and $B = \{1, 10, 100, 1000\}$
(b) $A = \{\text{red, blue, indigo}\}$ and $B = \{\text{orange, yellow}\}$
(c) $A = \{x: x \text{ is a real number and } 0 \le x < 1\,000\,000\}$

Solution:
(a) $S = \{x: x \text{ is a decimal number}\}$

(b) S = {c: c is a colour of the rainbow}

(c) A = {x: x is a positive real number}

Worked example 3.10

Describe the complements of each of the following:

(a) A = {2, 4, 6, 8, 10} where S = {x: x is an integer and $0 \leq x < 12$}

(b) A = {red, blue, green, yellow} where S = {x: x is a colour of the rain-
bow}

(c) A = {a, e, i, o, u} where S = {l: l is a lower-case letter of the alphabet}

Solution:

(a) A′ = {0, 1, 3, 5, 7, 9, 11}

(b) A′ = {orange, indigo, violet}

(c) A′ = {x: x is a consonant}

Worked example 3.11

List the contents of set Q where

(a) Q = (A ∩ B)′, where A = {2, 5, 6, 7, 9, 12}, B = {1, 3, 5, 7, 9, 10} and
S = {x: x is an integer and $0 \leq x < 15$}

(b) Q = (A′ ∪ B′), where A = {x: x is a vowel}, B = {x: x is a consonant}
and S = {x: x is a letter of the alphabet}

(c) Q = A′ ∪ (B ∩ C′), where A = {Jones, Sykes, Smith}, B = {Brown,
Taylor}, C = {Sykes, Brown, Murphy} and S = {Jones, Sykes, Smith,
Brown, Taylor, Murphy}

Solution:

(a) Q = (A ∩ B)′ = [{2, 5, 6, 7, 9, 12} ∩ {1, 3, 5, 7, 9, 10}]′

= {5, 7, 9}′

= {0, 1, 2, 3, 4, 6, 8, 10, 11, 12, 13, 14}

(b) Q = (A′ ∪ B′) = {vowel}′ ∪ {consonant}′

= {consonant} ∪ {vowel}

= S

(c) Q = A′ ∪ (B ∩ C′)

= {Jones, Sykes, Smith}′ ∪ [{Brown, Taylor} ∩ {Sykes, Brown,
Murphy}′]

= {Brown, Taylor, Murphy} ∪ [{Brown, Taylor} ∩ {Jones,
Smith, Taylor}]

= {Brown, Taylor, Murphy} ∪ [{Taylor}]

= {Brown, Taylor, Murphy}

Exercises

3.9 Define suitable universal sets for each of the following:
(a) A = {−5, −3, −1, 1, 3, 5} and B = {−4, −2, 0, 2, 4}
(b) A = {pen, paper, stapler, ruler} and B = {ink, eraser, pencil sharpener}
(c) A = {x: x is a prime number and $0 \leq x < 23$}

3.10 Describe the complements of each of the following:
(a) A = {1, 3, 5, 7, 9} where S = {x: x is a natural number and $0 < x < 10$}
(b) A = {a, c, e, f} where S = {a, b, c, d, e, f, g, h}
(c) A = {January, March, July, September, December} where S = {m: m is a month of the year}

3.11 List the contents of set Q where:
(a) Q = (A ∪ B)′, where A = {001, 010, 011, 101}, B = {001, 010, 100} and S = {x: x is a binary number and $000 \leq x \leq 111$}
(b) Q = (A′ ∩ B′), where A = {x: x is an even integer}, B = {x: x is an odd integer} and S = {x: x is an integer}
(c) Q = A′ ∩ (B ∪ C′), where A = {lemonade, tea, ginger beer}, B = {tea, orange juice}, C = {Coca-Cola, tea, coffee} and S = {lemonade, Coca-Cola, orange juice, tea, coffee, ginger beer}

Disjoint sets
If two sets have no elements in common they are said to be disjoint sets. For example, if A = {2, 4, 6} and B = {1, 3, 5} then

$$A \cap B = \varnothing$$

that is the sets are disjoint.

Set partitions
Every set can be constructed from a union of disjoint subsets. For example, the set

$$A = \{\text{spades, hearts, clubs, diamonds}\}$$

has, among the elements of the power set of A, the following disjoint subsets of A:

A1 = {hearts}
A2 = {spades, diamonds}
A3 = {clubs}

where

$$A = A1 \cup A2 \cup A3$$

The disjoint subsets A1, A2 and A3 form what is called a **partition** of the set A. Every set can be partitioned into a collection of disjoint subsets. Notice that the partition is not unique; the following is an alternative partition of A:

A4 = {spades}
A5 = {hearts}
A6 = {clubs, diamonds}

Conclusion: Every set can be partitioned into a collection of disjoint subsets.

Worked example 3.12

Which of the following are partitions of the set A?
(a) A1 = {1, 3, 7, 11}
     A2 = {2, 4, 8}
     A3 = {1, 5, 6, 9}
     A4 = {10}
     where A = {x: x is an integer and $0 \leq x \leq 10$}
(b) A1 = {red, yellow}
     A2 = {blue, green}
     A3 = {indigo, violet}
     where A = {x: x is a colour of the rainbow}
(c) A1 = {January, March, April}
     A2 = {December}
     A3 = {February, August, November}
     A4 = {May, July, September}
     A5 = {June, October}
     where A = {x: x is the name of a month}

Solution:
(a) This is not a partition because A1 and A3 are not disjoint, A1 $\cap$ A3 = {1}.
(b) This is not a partition because the union of A1, A2 and A3 is not equal to A. The colour orange is missing.
(c) This is a partition of A because each of the listed subsets of A is disjoint from every other listed subset of A and A can be constructed from their union.

Exercise

3.12 Which of the following are partitions of the set A?
   (a)   A1 = {−1, 2, −2}
         A2 = {3, −3}
         A3 = {0, 1, 2}
   where A = {x: x is an integer and −3 ≤ x ≤ 3}
   (b)   A1 = {Monday, Tuesday}
         A2 = {Saturday, Thursday, Sunday}
         A3 = {Friday}
   where A = {x: x is a day of the week}
   (c)   A1 = {0, 5, B}
         A2 = {E, 3}
         A3 = {1, 2, A, D}
         A4 = {F, 4, 9, 8}
         A5 = {6, 7, C}
   where A = {x: x is a hexadecimal number and 0 ≤ x ≤ F}

Diagrammatic representation

The Venn diagram

Whenever a mathematical concept is illustrated with a diagram, erstwhile complications often appear to evaporate. Diagrams convey large quantities of information for seemingly little content and such is the case with Venn diagrams – invented by John Venn to display properties of sets.

The diagrams are very simple. A rectangle is drawn, whose interior points are taken to represent the elements of the universal set (Figure 3.1).

S

Figure 3.1

Any subset of the universal set is represented by a circle drawn entirely within the confines of the rectangle, the elements of the subset being represented by the points inside the circle (Figure 3.2).

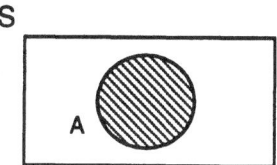

Figure 3.2

Each element of the universal set that is not an element of the subset is an element of the complement of the subset and as such is represented by a point outside the circle but inside the rectangle (Figure 3.3).

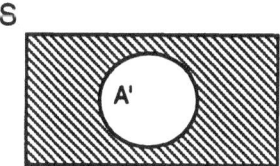

Figure 3.3

Union of two sets
Following the conventions of the Venn diagram for a set contained within the universal set we can easily display the concept of a union of two such sets as shown in Figure 3.4.

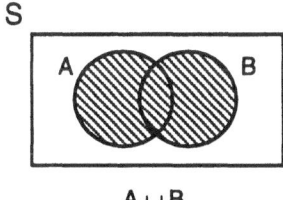

Figure 3.4

Intersection of two sets
The intersection of two sets consist of those elements that are common to both sets. This lends itself to the graphical illustration in Figure 3.5.

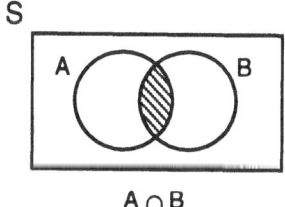

Figure 3.5

DeMorgan's laws

As has been stated, the Venn diagram provides a most useful graphical display of set properties. In particular, it can be used to demonstrate equivalences of certain expressions involving sets.

Two such equivalences, which have their counterparts in the algebra of propositions, are contained within DeMorgan's laws, one of which states that

$$(A \cap B)' = A' \cup B'$$

The complement of the union of two sets is the intersection of the complements of the two sets. To illustrate this property we use an appropriate Venn diagram (Figure 3.6).

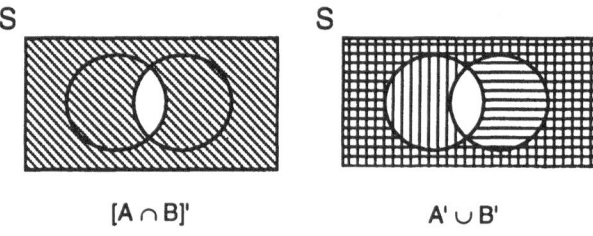

$[A \cap B]'$ $A' \cup B'$

Figure 3.6

Conclusion: A Venn diagram is a pictorial aid to the manipulation of sets.

Worked example 3.13

Draw the Venn diagram that corresponds to each of the following:
(a) $(A \cup B)'$
(b) $A \cup (B \cap C)$

Solution:
(a) See Figure 3.7.
(b) See Figure 3.8.

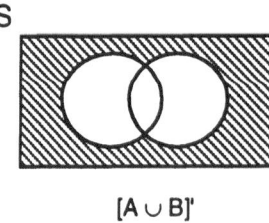

$[A \cup B]'$

Figure 3.7

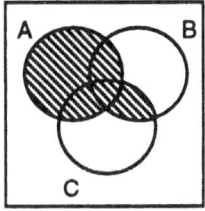

A ∪ [B ∩ C]

Figure 3.8

Worked example 3.14

Use a Venn diagram to illustrate the following equivalences:
(a) $(A \cup B)' = A' \cap B'$
(b) $A \cup (B \cap C) = (A \cup B) \cap (A \cup C)$

Solution:
(a)   See Figure 3.9.
(b)   See Figure 3.10.

S

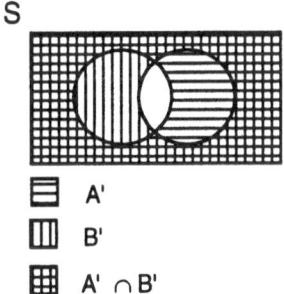

⊟   A'

�𝍇   B'

⊞   A' ∩ B'

Figure 3.9

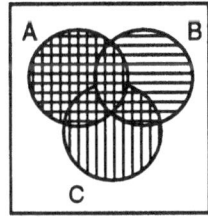

⊟   A ∪ B

⫿   A ∪ C

⊞   [A ∪ B] ∩ [A ∪ C]

Figure 3.10

Exercises

3.13 Draw the Venn diagram that corresponds to each of the following:
 (a) $A \cap (A \cup B)$
 (b) $A \cap (B \cup C)$

3.14 Use a Venn diagram to illustrate the following equivalences:
 (a) $A \cap (A \cup B) = A$
 (b) $A \cap (B \cup C) = (A \cap B) \cup (A \cap C)$

The algebra of sets

In Chapter 2 we saw that propositions and their associated logical connectives obeyed a number of laws or rules that were collectively referred to as the **algebra of propositions**. This is a general feature of mathematical constructs. Whenever a collection of mathematical objects and their associated operations obey a set of rules then the objects, the operations and the rules collectively form what is termed an **algebra**. The theory of sets is no exception. Here the mathematical objects are sets and their associated operations are union, intersection and complement. What follows is the set of rules that are obeyed and which couples with sets and their operations to form the **algebra of sets**.

Commutativity and associativity

We have already seen that union and intersection are both commutative and associative:

$$A \cup B = B \cup A \text{ and } A \cap B = B \cap A$$

$$(A \cup B) \cup C = A \cup (B \cup C) = A \cup B \cup C \text{ and}$$

$$(A \cap B) \cap C = A \cap (B \cap C) = A \cap B \cap C$$

Distributivity

The operation of union is distributive over intersection and the operation of intersection is distributive over union:

$$A \cup (B \cap C) = (A \cup B) \cap (A \cup C) \text{ and}$$

$$A \cap (B \cup C) = (A \cap B) \cup (A \cap C)$$

These two rules can be easily demonstrated using a Venn diagram (Figure 3.11).

Notice that this demonstration does not prove that the distribution law holds. To prove that the law is true requires a more detailed discussion of

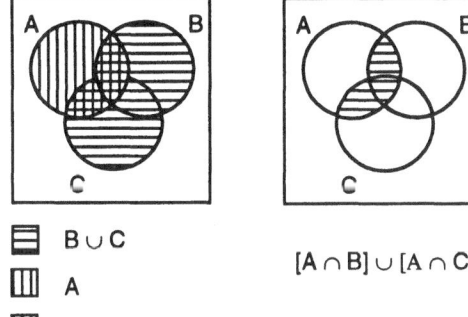

$B \cup C$

A

$A \cap [B \cup C]$

$[A \cap B] \cup [A \cap C]$

Figure 3.11

sets than is appropriate in this book. The Venn diagram is sufficient to demonstrate its validity for that particular arrangement of sets.

Identity
We have already seen that

$$A \cup S = S \text{ and } A \cup \varnothing = A$$

and that

$$A \cap S = A \text{ and } A \cap \varnothing = \varnothing$$

Idempotency
From the definition of a set it is clear that

$$A \cup A = A \text{ and } A \cap A = A$$

Negations
The union of a set with its complement is the universal set and the intersection of a set with its complement is the universal set

$$A \cup A' = S \text{ and } A \cap A' = \varnothing$$

DeMorgan's laws
Again we have already seen that

$$(A \cup B)' = A' \cap B' \text{ and } (A \cap B)' = A' \cup B'$$

Involution
The complement of a complement of a set is the set itself:

$(A')' = A$

Absorption

$A \cup (A \cap B) = A$ **and** $A \cap (A \cup B) = A$

These two rules can be easily demonstrated using Venn diagrams (see Exercise 3.14).

Conclusion: Sets and their associated operations of union, intersection and complement couple with a collection of rules that collectively define the algebra of sets.

Worked example 3.15

Using a Venn diagram, demonstrate the equivalence of each of the following pairs of expressions:
(a) $(A \cap B') \cup (A' \cap C) \cup (A \cap B) = A \cup C$
(b) $(A \cap B \cap C) \cup (A \cap B \cap C') = A \cap B$

Solution:
(a) See Figure 3.12.
(b) See Figure 3.13.

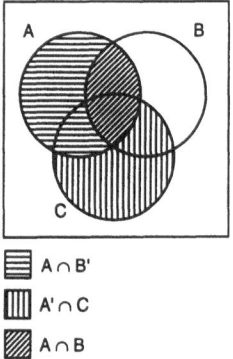

▤ $A \cap B'$
▥ $A' \cap C$
▨ $A \cap B$

Figure 3.12

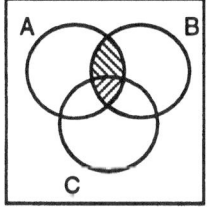

■ $A \cap B \cap C'$
▨ $A \cap B \cap C$

Figure 3.13

Exercise

3.15 Using a Venn diagram, demonstrate the equivalence of each of the following pairs of expressions:
(a) $(A \cap C) \cup (A \cap C') \cup (B' \cap C) = A \cup (B' \cap C)$
(b) $(A \cap B' \cap C) \cup (A \cap B' \cap C') \cup (A \cap B \cap C) = A \cap (B' \cup C)$

Set operations and the logical connectives
By now you will no doubt be experiencing *déjà vu* – you have seen these laws before. Indeed, the laws of the algebra of sets and the laws of the algebra of propositions (and of switching circuits for that matter) have the same names and very similar appearances. In fact, replacing sets with propositions, union with conjunction, intersection with disjunction and complement with negation transforms the algebra of sets into the algebra of propositions. This similarity is said to be an **isomorphism** – an exact correspondence in form and relations.

The reason for this isomorphism between sets and propositions is:

(a) A union of two sets implies a conjunction:

$A \cup B = \{x: (x \in A) \vee (x \in B)\}$

(b) An intersection of two sets implies a disjunction:

$A \cap B = \{x: (x \in A) \wedge (x \in B)\}$

(c) A complement of a set implies a negation:

$A' = \{x: (x \notin A)\}$ **or** $A' = \{x: \neg(x \in A)\}$

The operations of union, intersection and complementation of sets obey the identical laws within set theory to those obeyed by the logical connectives of OR, AND and NOT within the propositional calculus. If we use membership of a set to correspond to the truth value of a proposition then we can see the similarities between the two branches of mathematics.

Conclusion: Sets and propositions are isomorphic to one another.

Worked example 3.16

Describe the following using propositions that describe membership of the component sets:
(a) $A \cap (B' \cup C)$
(b) $(A' \cup B)'$

Solution:
(a) $\{x: (x \in A) \wedge [\neg(x \in B) \vee (x \in C)]\}$
(b) $\{x: \neg[\neg(x \in A) \vee (x \in B)]\}$

Worked example 3.17

With the aid of a Venn diagram, describe the sets that are implied by the following compound propositions:
(a) $\{(x \in A) \vee (x \in B)\} \wedge (x \in A)$
(b) $(x \in A') \wedge \{(x \in B) \vee (x \in C')\}$
(c) $\{\neg(x \in A)\} \vee (x \notin B)$

Solution:
(a) See Figure 3.14.
(b) See Figure 3.15.
(c) See Figure 3.16.

S

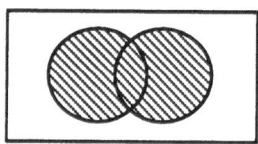

   $(x \in A \vee x \in B) \wedge x \in A$

$A \cup B$

Figure 3.14

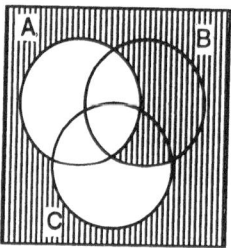

   $A' \cap (B \cup C')$

Figure 3.15

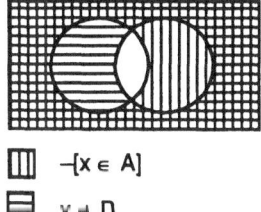

▥ ¬[x ∈ A]

▤ x ∉ D

▦ ¬[x ∈ A] ∧ [x ∉ B]

Figure 3.16

Exercises

3.16 Describe the following sets using propositions that describe membership of the component sets:
  (a) $A' \cup (B \cap C')$
  (b) $A' \cap B'$

3.17 With the aid of a Venn diagram, describe the sets that are implied by the following compound propositions:
  (a) $\neg\{(x \in A) \land (x \notin B)\}$
  (b) $(x \in A) \lor \{(x \in B) \land \neg(x \in C)\}$
  (c) $\{\neg(x \in A)\} \land (x \in B')$

The exclusive OR

We saw that in the propositional calculus the logical connective OR was an inclusive OR – it permitted that truth of both propositions that it connected. We also saw that in the English language there is some ambiguity as to whether OR means **either one or the other or both** or whether it means **either one or the other but not both**. This latter meaning is the exclusive OR, abbreviated to EOR. In set theory, this has its counterpart in the **symmetric difference**.

The difference

We define the difference between set A and set B as the set containing those elements that are in A that are not in B (Figure 3.17). The symbolism that we use for the difference between two sets is

  $A - B$

where it can easily be seen that

  $A - B \equiv A \cap B'$

S

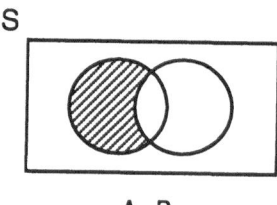

A - B

Figure 3.17

The symmetric difference

The symmetric difference between two sets is defined in Figure 3.18.

S

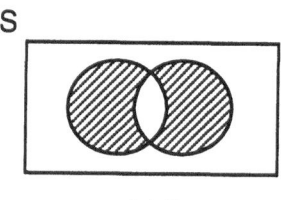

A △ B

Figure 3.18

$$A \triangle B = (A - B) \cup (B - A) \equiv (A \cap B') \cup (B \cap A')$$

From this it can be seen that in the propositional calculus the exclusive OR (EOR) is:

$$p \veebar q \equiv (p \wedge \neg q) \vee (q \wedge \neg p)$$

where $\veebar$ stands for the EOR logical connective.

Conclusion: The set operation symmetric difference has its counterpart in the propositional calculus in the exclusive OR.

Worked example 3.18

Show that:
(a) $(A \triangle B) \triangle C = A \triangle (B \triangle C)$
(b) $A \cup (B \triangle C) \neq (A \cup B) \triangle (A \cup C)$

Solution:
(a) See Figure 3.19.
(b) See Figure 3.20.

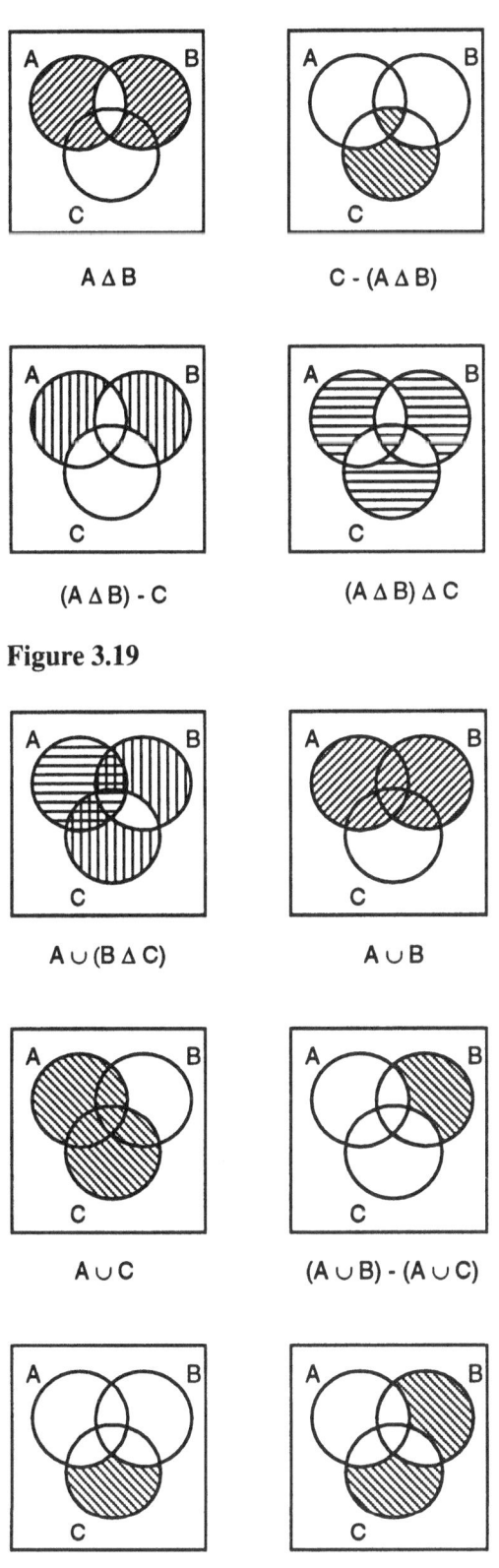

Figure 3.19

Figure 3.20

Exercise

3.18 Show that:
  (a) $A' \Delta B' \neq (A \Delta B)'$
  (b) $A \cap (B \Delta C) = (A \cap B) \Delta (A \cap C)$

Database files

A database file is a set of records, each of which contains information set against a regular format. This record format, or template, is in the form of a collection of named fields, and information is entered into the record by giving values to each field. For example, the following represents a typical record from a database file of student details:

Student File

Record Number: 147   **Enrolment Date**: 04/09/93
Name: A Student   **Student Number**: 123456
School: Science   **Department**: Physics
Course: BSc (Applied Acoustics)   **Year**: 2

At the top of each record is the name of the file. Within each record are the field names, separated from their values in that particular record by colons.

Querying a database file

A database file stores information in a format that is designed for ease of access, and the process of accessing information is called **querying** the file. For example, if you wished to see the names of every student in the Student File who enrolled in 1994 you would query the file with a query statement of the form:

  [Enrolment Date = ??/??/94]

This query will cause the computer to range through the file and select every record that has the number 94 as the year in the Enrolment Date field – the question mark (?) is called a wild card and permits any single character to be substituted.

  When the query has been executed all the selected records will be stored in a query file. If the original database file is thought of as the universal set then any query file is a subset of the universal set:

  S = {x: x is a Student File record}

  Q = {x: x is a Student File record AND record Enrolment Date = ??/??/94}

S

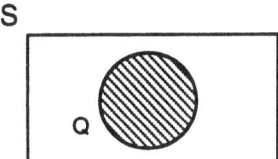

Figure 3.21

Conclusion: A database file is a universal set of records, subsets of which can be created by querying the database. Queries are in the form of compound propositions.

Worked example 3.19

A database file has a record template containing the following fields:

Name
Last Month Sales
Balance Due
Credit Rating

where the value of Credit Rating is 0, 1, 2 or 3.

Describe the set of records created as a result of executing each of the following queries:
(a) (Balance Due > 1000) ∧ (Credit Rating = 1)
(b) (Last Month Sales > 2500) ∧ (Credit Rating = 0)) ∨ ((Last Month Sales ≤ 2500) ∧ (Credit Rating < 2))
(c) (Name = A*) ∧ (Credit Rating = 3)
     where * is a wild card that can be substituted by any string of characters.

Solution:
(a) $A \cap B$ where $A = \{x: x$ record with Balance Due > 1000$\}$ and $B = \{x: x$ record with Credit Rating = 1$\}$
(b) $(A \cap B) \cup (A' \cap C)$ where $A = \{x: x$ record with Last Month Sales > 2500$\}$, $B = \{x: x$ record with Credit Rating = 0$\}$ and $C = \{x: x$ record with Credit Rating < 2$\}$
(c) $A \cap B$ where $A = \{x: x$ record with Name beginning with the letter A$\}$ and $B = \{x: x$ record with Credit Rating = 3$\}$

Worked example 3.20

Describe the queries necessary to create the following sets of records from the database file of the previous example, where A = {Names beginning

with C}, B = {Last Month Sales < 2000}, C = {Balance Due ≥ 1000} and
D = {Credit Rating ≥ 2}:
(a) A ∩ (B ∩ C)
(b) C ∩ (D ∪ C)
(c) A ∪ C′

Solution:
(a) (Name = C*) ∧ ((Last Month Sales < 2000) ∧ (Balance Due ≥ 1000))
(b) (Balance Due ≥ 1000) ∧ ((Credit Rating ≥ 2) ∨ (Balance Due ≥ 1000))
(c) (Name = C*) ∨ (Balance Due < 1000)

Exercises

3.19 A database file has a record template containing the following fields:

Name
Best Score
Worst Score
Last Score

Describe the set of records created as a result of executing each of the
following queries:
(a) (Worst Score > 100) ∧ (Best Score < 80)
(b) (Name = B*) ∧ ((Worst Score < 95) ∨ (Best Score ≥ 76))
(c) ((Best Score < 100) ∧ (Worst Score > 100)) ∧ Last Score < 100)

3.20 Describe the queries necessary to create the following sets of records
from the database file of the previous exercise, where A = {Names
beginning with R}, B = {Worst Score < 105}, C ={Best Score ≥ 85}
and D = {Latest Score ≥ 92}:
(a) A ∪ (B ∩ C)
(b) C ∪ (D ∩ C)
(c) A ∩ D′

Numbers

Computing and mathematics
During the industrial revolution, processes that hitherto were performed by
manual labour were transferred to machines. Invention followed desire.
James Hargreaves' spinning-jenny, Samuel Compton's mule and the rotary-
motion steam engine invented by James Watt are all machines that were
developed from a desire to **mechanize**. The situation today is little different.
The desire to **computerize** is driving the computer revolution, and increas-
ingly more systems are being placed under the control of a computer.
Where the situation does differ is in the way the desire is implemented. We

no longer look at a task and design a computer – a machine – to perform it. Instead we look at a task and design a piece of software to drive a ready-made computer to perform the task. When the hardware does not match the needs of the software we develop and extend the hardware; so often the creation and needs of software drive the development of the hardware.

Before software is implemented on a system it must be proved to be correct – too many delicate systems require this for it to be otherwise: from the control of credit within a commercial organization to the control of life-threatening radiation in a nuclear power station. For a small system it may be possible to test the software with trial data and thereby arrive at a conclusion that the software is performing correctly and so is correct beyond reasonable doubt. This, however, is not the same as **proving** that it is correct – it has only demonstrated that it is correct for the trial data used. The larger software systems become, the less adequate this method of testing proves to be. For all systems there is a crucial need to prove the software correct beyond any doubt whatsoever and not just beyond reasonable doubt.

This need to prove program correctness has been the stimulus behind the use of mathematical concepts in software development. Mathematics is exact; there is no ambiguity. Mathematics is universal; there is no contention. Mathematics is analytic; it possess the rigorous framework against which correct software can be developed and against which software can be proved to be correct.

Mathematics concerns itself with the commonality of different ideas and attempts to express this commonality with as few basic facts as possible. The propositional calculus and switching circuits are two distinct ideas. Set theory is a third idea, but to the mathematician they are all just exemplar illustrations of a single idea – the idea of a Boolean algebra. It is not appropriate to discuss Boolean algebra here; instead it is sufficient to mention that we have seen and discussed the commonality in the propositional calculus, switching circuits and set theory.

The search for commonality and the reduction of that commonality to its simplest basics permits complicated mathematical superstructures to be constructed in the knowledge that because the foundation is firm the edifice will stand. And so it is with software. We must be certain that the fundamentals are solidly based before we can create the superstructure, and in what follows we shall glimpse an indication of how structures can be constructed from simple beginnings with the aim of demonstrating both this principle and the important process of **recursion**.

Numbers and counting

Many centuries ago records of quantities were kept by using a tally system. In the Musée de L'Homme in Paris there is an ancient Peruvian Indian census *quipu* that consists of a collection of cords, each cord containing a number of knots with large knots representing multiples of small knots. The knots are effectively records of quantities with different coloured cords indicating male and female. This was the tally system. There was no notion of number as an abstract idea, the quantity of one item was recorded by the quantity of another item.

When *Homo sapiens* changed from being a hunter–gatherer to a community-based farmer, social order demanded that counting become more intensive and a need arose for a more systematic counting system. The Babylonians used wedge-shaped grooves cut into a wax tablet for their tally system. When quantities became too large to be accommodated on a single wax tablet then it would have become apparent that the tablet could be more efficiently used if different marks were used for different numbers. From this realization it is but a short step to the development of a fully fledged number system: an abstract system of numbers that could be used to count men and women in a census, count knots on a string and even count numbers themselves.

Counting is perhaps the simplest abstract mathematical process known to mankind – the association of quantities of things with the abstract notion of number. To count we need to understand the nature of numbers and how they combine under the various arithmetic operations. The computer needs to know all this as well, which takes us back to the beginning of this chapter where we discussed data types.

Data types are fundamental to good programming practice, and in programming languages such as Pascal, Modula 2 and C, certain data types are defined by default. Typical of default sets of numerical data type values are:

The natural numbers:

$$\mathcal{N} = \{x: x \text{ is a natural number}\}$$

A natural number is a positive, whole number or zero, signified by the numerals 0, 1, 2, 3…

The integers:

$$Z = \{x: x \text{ is an integer}\}$$

An integer is a positive or negative whole number where negative whole numbers are indicated by being prefixed with a minus sign $(-)$.

The rationals:

$$\mathcal{Q} = \{x: x \text{ is a rational number}\}$$

A rational number is a ratio of whole numbers in the form a/b where a, $b \in Z$ and $b \neq 0$.

The reals:

$$\mathcal{R} = \{x: x \text{ is a real number}\}$$

The definition of a real number requires a more detailed consideration of the rational numbers.

Each of these sets of data type values is coupled with the arithmetic operations of addition, subtraction, multiplication, division and raising to a power to complete the definitions of the data types.

Of these four data types the last three can be constructed from the first, the natural numbers. Taking this view, the natural numbers form what is called an atomic data type and the other three each form a data structure, which is a data type that is constructed from a more primitive data type. Here then is where we must start: the data type of **natural numbers** which will form the basis for the entire number system.

The natural numbers can be taken to be intuitively accepted. After all, we all know what a whole number is; we use them every day. Three coins in one fountain, two turtle doves and one partridge in one pear tree. Indeed, earlier in this chapter the cardinality of a set was defined as the number of elements that it contained. This, however, assumed that the concept of natural number had been explained when in fact the opposite was true.

At the same time we intuitively accept the notion of addition. We know that two plus two equals four: we were told that when we were just so high. We also know how to multiply – that is, those of us who spent time learning our times tables know how to multiply. So should we leave our notions of natural number there? Or is there a more appropriate set of basics from which we can derive **all** of these data types, more appropriate, that is, to the task of constructing data types? The answer to the last question is that there is.

We are in the business of constructing data types, and all data types list their data type values in a set. In other words, the notion of a set is common to all data types and so we shall take as our most primitive data type values, the set of sets.

The natural numbers

From the atomic data type of set we can construct the natural numbers. The simplest set possible is the empty set $\varnothing$ and this, by definition, we equate with the natural number zero:

$$0 = \varnothing$$

Zero is defined to be the empty set. This is the fundamental definition. We call it an **axiom**; a basic, self-evident truth which is not based upon any previous fact. Notice that zero is defined as a set; we shall say more about this later.

We now define unity to be the set that contains the empty set:

$$1 = \{\varnothing\} = \{0\}$$

from the definition of 0. Notice that this set contains one element – it has cardinality 1.

Two is defined to be the set that contains the empty set and the set that contains the empty set:

$$2 = \{\varnothing, \{\varnothing\}\} = \{0, 1\}$$

from the definitions of 0 and 1. Notice that this set contains two elements – it has cardinality 2.

Repeating this procedure we define three to be the set that contains the empty set, the set that contains the empty set and the set that contains the empty set and the set that contains the empty set:

$$3 = \{\varnothing, \{\varnothing\}, \{\varnothing, \{\varnothing\}\}\} = \{0, 1, 2\}$$

from the definitions of 0, 1 and 2. Again, this set contains three elements – it has cardinality 3.

We can continue in this way indefinitely where each new natural number is defined in terms of the previously defined natural number. One important point revealed by this process is that any natural number can be defined in terms of the original axiom:

$$0 = \varnothing$$

by continually repeating the same sequence of actions.

Recursive processes

A **recursive process** is one in which new results are derived from old results by continually repeating the same sequence of actions. The construction of the natural numbers just displayed is an example of a recursive process; by use of the set that represents a given natural number the next natural number can be obtained. By continually repeating this sequence of actions indefinitely, the natural numbers can be defined.

Conclusion: The data type values of natural numbers can be constructed from the atomic data type of sets. This is achieved using a recursive process in which the set that represents a natural number is used to define the next natural number.

Successor and predecessor

Notice how we are building up these numbers. They are increasing in value by unity at a time and each new number, with the exception of 0, is the union of the previous number with the set containing the previous number:

$$1 = 0 \cup \{0\} = \varnothing \cup \{\varnothing\} = \{\varnothing\}$$

$$2 = 1 \cup \{1\} = \{\varnothing\} \cup \{\{\varnothing\}\} = \{\varnothing, \{\varnothing\}\}$$
$$3 = 2 \cup \{2\} = \{\varnothing, \{\varnothing\}\} \cup \{\{\varnothing, \{\varnothing\}\}\} = \{\varnothing, \{\varnothing\}, \{\varnothing, \{\varnothing\}\}\}$$

Notice that the process that is being used here is a **recursive** process because it repetitively uses the previous definition to define the next number.

In general we can define the next natural number after n as:

$$n \cup \{n\}$$

This idea of the next natural number is basic. Every natural number has a next natural number that is larger by unity. This is referred to as the **ordering** of the natural numbers. The process of obtaining the next natural number we shall define as an operation SUC on the last natural number obtained:

$$\text{SUC } n = n \cup \{n\}$$

where SUC n stands for the **successor to** n, the successor to n being the union of n with the set whose only element is n.

Clearly, if every natural number has a successor it also has a **predecessor**, except, that is, if the number is zero. We shall define the process of finding the predecessor using an operator PRE where

$$n = \text{PRE (SUC } n)$$

The natural number n is the predecessor of the successor to itself.

Thus far we have defined the natural numbers as sets and we have embodied the concept of the order of the natural numbers in the definitions of the successor and the predecessor to a number. Next we must consider the process of adding one natural number to another.

Note that
$n = $ SUC (PRE n) does not work because we have not yet defined PRE 0.

Addition

The process of adding two numbers together we shall denote by the operation ADD, which is defined as:

$$n \text{ ADD } m = m \qquad \text{if } n = 0$$
$$= \text{SUC ([PRE } n] \text{ ADD } m) \qquad \text{otherwise}$$

This may seem to be a rather complicated combination of operations but it is recursive and simple to use. For example

$$0 \text{ ADD } 3 = 3 \quad \text{from the first line of the definition}$$

and

2 ADD 2	= SUC ([PRE 2] ADD 2)	from the second line of the definition
	= SUC (1 ADD 2)	1 is the predecessor of 2
	= SUC (SUC [(PRE 1) ADD 2])	from the second line of the definition
	= SUC (SUC [0 ADD 2])	0 is the predecessor of 1
	= SUC (SUC 2)	from the first line of the definition
	= SUC 3	3 is the successor of 2
	= 4	4 is the successor of 3

Here then we have defined addition without using addition itself. Having done this we now define the symbol + to stand for the operation ADD whence we find that

$$0 + 3 = 3 \text{ and } 2 + 2 = 4$$

In this way we can define the addition of any pair of natural numbers. It does not matter that we have defined the natural numbers as sets because the arithmetic process of adding is one of manipulating symbols and we have just shown how to manipulate the symbols $0 + 3$ to arrive at the symbol 3 and the symbols $2 + 2$ to arrive at the symbol 4. The same can be said of any arithmetic process: it is a manipulation of symbols. This is what the Peruvian census taker did – manipulate symbols in the form of knots.

Conclusion: The arithmetic process of addition can be defined in terms of the successor and predecessor operations. The outcome is a fully consistent method of manipulating symbols in the form of the numerals and the arithmetic operation +.

Worked example 3.21

Show, by considering the set that defines the natural number 3, that the number 4 can be given as $\{0, 1, 2, 3\}$.

Solution: $4 = \{\varnothing, \{\varnothing\}, \{\varnothing, \{\varnothing\}\}, \{\varnothing, \{\varnothing\}, \{\varnothing, \{\varnothing\}\}\}\} = \{0, 1, 2, 3\}$.

Worked example 3.22

Use the successor and predecessor operations to demonstrate that $3 + 1 = 4$.

Solution:
3 ADD 1 = SUC ([PRE 3] ADD 1)

$$= \text{SUC} (2 \text{ ADD } 1)$$
$$= \text{SUC} (\text{SUC} [(\text{PRE } 2) \text{ ADD } 1])$$
$$= \text{SUC} (\text{SUC} [1 \text{ ADD } 1])$$
$$= \text{SUC} (\text{SUC} [\text{SUC} ([\text{PRE } 1] \text{ ADD } 1)])$$
$$= \text{SUC} (\text{SUC} [\text{SUC} (0 \text{ ADD } 1)])$$
$$= \text{SUC} (\text{SUC} [\text{SUC } 1])$$
$$= \text{SUC} (\text{SUC } 2)$$
$$= \text{SUC } 3$$
$$= 4$$

Therefore $3 + 1 = 4$.

Exercises

3.21 Show, by considering the set that defines the natural number 4, that the number 5 can be given as $\{0, 1, 2, 3, 4\}$.

3.22 Use the successor and predecessor operations to demonstrate that $2 + 3 = 5$.

Subtraction

Subtraction is defined as the reverse operation to addition. The process of subtracting one number from another we shall denote by the operator SUB, which is defined as

$$n \text{ SUB } m \ = \ 0 \qquad\qquad\qquad \text{if } n = m$$
$$= \ \text{SUC} ([\text{PRE } n] \text{ SUB } m) \qquad \text{otherwise}$$

Hence, for example

$3 \text{ SUB } 3 = 0$ from the first line of the definition

and

$5 \text{ SUB } 3 = \text{SUC} ([\text{PRE } 5] \text{ SUB } 3)$	from the second line of the definition
$= \text{SUC} (4 \text{ SUB } 3)$	4 is the predecessor of 5
$= \text{SUC} (\text{SUC} [(\text{PRE } 4) \text{ SUB } 3])$	from the second line of the definition
$= \text{SUC} (\text{SUC} [3 \text{ SUB } 3])$	3 is the predecessor of 4
$= \text{SUC} (\text{SUC } 0)$	from the first line of the definition
$= \text{SUC } 1$	1 is the successor of 0
$= 2$	2 is the successor of 1

This definition of SUB raises problems when we attempt to take a larger number from a smaller. For example

1 SUB 2 = SUC ([PRE 1] SUB 2)	from the second line of the definition
= SUC (0 SUB 2)	0 is the predecessor of 1
= SUC (SUC [(PRE 0) SUB 2])	from the second line of the definition

and here is our problem: what is PRE 0? We do not have a number that is a predecessor to zero.

Do we now accept that we can only subtract a greater number from a lesser number or do we look for a way around the problem? We look for a way around the problem and resolve it by defining negative whole numbers as natural numbers prefixed with a minus sign $(-)$. This permits the extension of the definition of the SUB operator as

n SUB m = 0	if $n = m$
= SUC ([PRE n] SUB m)	if $n > m$
= $-(m$ SUB $n)$	if $n < m$

In this case, for example

3 SUB 5 = $-(5$ SUB 3)	because $3 < 5$
= -2	

Having resolved the problem of subtracting a larger number from a smaller number we have had to extend the set of data values to include negative whole numbers. The union of the natural numbers with the negative whole numbers forms the **integers**. Hence the data type Z has been constructed from the data type $\mathcal{N}$.

Conclusion: The arithmetic process of subtraction can be defined in terms of the successor and predecessor operations. To produce an operation that has no restriction on the natural numbers that are subtracted from each other we must extend the natural numbers to the integers.

Worked example 3.23

Show, using the successor and predecessor operations, that
(a) $5 - 2 = 3$
(b) $4 - 3 = 1$
(c) $3 - 4 = -1$

Solution:

(a) 5 SUB 2 = SUC ([PRE 5] SUB 2)
　　　　　　 = SUC (4 SUB 2)
　　　　　　 = SUC (SUC [(PRE 4) SUB 2])
　　　　　　 = SUC (SUC [3 SUB 2])
　　　　　　 = SUC (SUC [SUC ([PRE 3] SUB 2)])
　　　　　　 = SUC (SUC [SUC (2 SUB 2)])
　　　　　　 = SUC (SUC [SUC 0])
　　　　　　 = SUC (SUC 1)
　　　　　　 = SUC 2
　　　　　　 = 3

(b) 4 SUB 3 = SUC ([PRE 4] SUB 3)
　　　　　　 = SUC (3 SUB 3)
　　　　　　 = SUC 0
　　　　　　 = 1

(c) 3 SUB 4 = −[4 SUB 3]
　　　　　　 = −1

Exercise

3.23 Show, using the successor and predecessor operations, that
　　　(a) $5 - 4 = 1$
　　　(b) $3 - 1 = 2$
　　　(c) $2 - 5 = -3$

Multiplication and division

Multiplication is defined as repetitive addition. For example

　$2 + 2 + 2 = 2 \times 3$

Division is defined as repetitive subtraction – the reverse process to multiplication. For example

　$6 - 2 - 2 - 2 = 0$ **or** 2 can be taken from 6 three times

This permits us to define the division operator ÷ as

　$6 \div 2 = 3$

The symbol ÷ can be replaced by / to form a ratio of integers:

　$6/2 = 3$

Again, we run into a problem. We now have ratios of integers but as yet no meaning to a ratio of the form, for example

$1/2$

To clear this problem by permitting **all** ratios of integers we thereby create numbers that are not themselves integers. In this way we extend the number system to Q – the **rational** numbers. Note that the use of the word rational indicates that they are **ratios** and has no connection with any other English use of the word. Notice the problem now with multiplication:

$(2/3) \times (4/5)$

It is meaningless to say this is repetitive addition – we cannot add 2/3 to itself 4/5ths of a time. Instead we rewrite the product as

$(2/3) \times (4/5) = (2 \times 4)/(3 \times 5) = 8/15$

and the problem is resolved.

Conclusion: The arithmetic process of multiplication is defined as repetitive addition and the process of division as repetitive subtraction. To produce an operation that has no restriction on the integers that are divided by each other we must extend the natural numbers to the rational numbers.

Worked example 3.24

Multiplication of x with y can be defined using the operator MUL as

$$x \text{ MUL } y = 0 \qquad\qquad\qquad \text{if } y = 0$$
$$\qquad\quad = x \text{ ADD } [x \text{ MUL } (\text{PRE } y)] \qquad \text{otherwise}$$

Show that:
(a) $2 \times 1 = 2$
(b) $2 \times 3 = 6$

Solution:
(a) 2 MUL 1 = 2 ADD [2 MUL (PRE 1)]
                 = 2 ADD [2 MUL 0]
                 = 2 ADD 0
                 = SUC ([PRE 2] ADD 0)
                 = SUC (1 ADD 0)
                 = SUC (SUC [(PRE 1) ADD 0])
                 = SUC (SUC [0 ADD 0])

$$
\begin{aligned}
&= \text{SUC (SUC 0)}\\
&= \text{SUC 1}\\
&= 2
\end{aligned}
$$

(b) 2 MUL 3 = 2 ADD [2 MUL (PRE 3)]

$$
\begin{aligned}
&= \text{2 ADD [2 MUL 2]}\\
&= \text{2 ADD [2 ADD (2 MUL [PRE 2])]}\\
&= \text{2 ADD [2 ADD (2 MUL 1)]}\\
&= \text{2 ADD [2 ADD (2 ADD [2 MUL (PRE 1)])]}\\
&= \text{2 ADD [2 ADD (2 ADD [2 MUL 0])]}\\
&= \text{2 ADD [2 ADD (2 ADD 0)]}\\
&= \text{2 ADD [2 ADD (SUC [(PRE 2) ADD 0])]}\\
&= \text{2 ADD [2 ADD (SUC [1 ADD 0])]}\\
&= \text{2 ADD [2 ADD (SUC [(SUC [(PRE 1) ADD 0])])]}\\
&= \text{2 ADD [2 ADD (SUC [(SUC [0 ADD 0])])]}\\
&= \text{2 ADD [2 ADD (SUC [SUC 0])]}\\
&= \text{2 ADD [2 ADD (SUC 1)]}\\
&= \text{2 ADD [2 ADD 2]}\\
&= \text{2 ADD 4}\\
&= 6
\end{aligned}
$$

Worked example 3.25

Division of integer x by integer y can be defined using the operator DIV as:

$$
\begin{aligned}
x \text{ DIV } y \; &= \; 1 && \text{if } x = y\\
&= \; [(x \text{ SUB } y) \text{ DIV } y] \text{ ADD } 1 && \text{if } x > y\\
&= \; x/y && \text{if } x < y
\end{aligned}
$$

Show that $4 \div 2 = 2$.

Solution:
$$
\begin{aligned}
\text{4 DIV 2} &= \text{[(4 SUB 2) DIV 2] ADD 1}\\
&= \text{[2 DIV 2] ADD 1}\\
&= \text{[([2 SUB 2] DIV 2) ADD 1] ADD 1}\\
&= \text{[0 ADD 1] ADD 1}\\
&= \text{1 ADD 1}\\
&= 2
\end{aligned}
$$

Exercises

3.24 Multiplication of x with y can be defined using the operator MUL as

$$
\begin{aligned}
x \text{ MUL } y \; &= \; 0 && \text{if } y = 0\\
&= \; x \text{ ADD } [x \text{ MUL (PRE } y)] && \text{otherwise}
\end{aligned}
$$

Show that:
(a) $3 \times 3 = 9$
(b) $4 \times 2 = 8$

3.25 Division of integer x by integer y can be defined using the operator DIV as

$$x \text{ DIV } y = 1 \qquad\qquad\qquad\qquad\quad \text{if } x = y$$
$$= [(x \text{ SUB } y) \text{ DIV } y] \text{ ADD } 1 \quad \text{if } x > y$$
$$= x/y \qquad\qquad\qquad\qquad\quad \text{if } x < y$$

Show that $6 \div 3 = 2$.

Raising to a power
The raising of a number to a power is defined as repetitive multiplication. For example

$$3 \times 3 \times 3 \times 3 = 3^4$$

This, however, only caters for positive natural number powers. To extend the definition to rational number powers we require further definitions:

$5^0 = 1$ Any number raised to a power zero is unity
$7^{-1} = 1/7$ A negative power indicates the reciprocal

Powers can be added, subtracted and multiplied:

$$2^3 \times 2^4 = 2^{3+4} = 2^7$$
$$2^3 \div 2^4 = 2^{3-4} = 2^{-1}$$
$$(2^3)^4 = 2^{3\times4} = 2^{12}$$

Division of powers causes a problem:

$$4^{6/3} = 4^2$$

this is no problem, but

$$2^{1/2}$$

is a problem.
    We define a number x raised to the power $1/n$ where $n \in \mathcal{N}$ as the nth root of x – that number which when multiplied by itself n times yields x. For example $5^{1/3}$ is the third root of 5 because

$$5^{1/3} \times 5^{1/3} \times 5^{1/3} = 5^{1/3+1/3+1/3} = 5^1 = 5$$

However, taking roots of a rational number sometimes produces numbers that are not themselves rational numbers. For example, $2^{1/2}$ is not a rational number. We shall leave the proof of this statement until Chapter 6, where we discuss arguments and the elements of proof. For now, we must accept that there are numbers that cannot be expressed as a ratio of two integers – we call them **irrational** numbers, where the word irrational means that it cannot be written as a ratio of two integers. Note that the use of the word irrational has no connection with any other use of the word in English.

The real numbers

The irrational numbers combine with the rational numbers to form the real numbers. Now our numbers are complete, that is with two provisos:

Division by zero is not defined

This is a restriction imposed on the real number system we must live with. There is no way we can define what is meant by division by zero.

The square root of a negative number is not a real number

This is a restriction imposed on the real number system that can be overcome. However, in overcoming it we must define another type of number, and that is another story altogether.

Conclusion: The arithmetic process of raising to a power is defined as repetitive multiplication. To produce an operation that has no restriction on the powers used we must extend the rational to the real numbers – the union of the rational numbers and the irrational numbers.

While it may seem that the real numbers are now completely defined we have not defined how to combine two irrational numbers under any one of the arithmetic operations.

Worked example 3.26

The assignment statement

$$x := 1 + y/(1 + x)$$

where x and y are real numbers, can be used recursively within a computer program. As the number of recursions increase the successive values of x approaches $(y + 1)^{1/2}$.

Use this assignment statement within a recursive for... to... loop to evaluate $\sqrt{2}$ after

(a) 5 recursions;
(b) 15 recursions;
(c) 100 recursions.

Solution:

```
program Irrational1(output);
var x : real; n : integer;
begin
        (*assign a starting value to x*)
        x := 1;
        (*set up a recursive loop to repeat 15 times*)
        for n := 1 to 15
          DO x := 1 + 1/(1 + x);
        writeln (x)
end.
```

By changing the number of times the recursive loop repeats you can obtain:
(a) 1.4141414141
(b) 1.4142135624
(c) 1.4142135624

Worked example 3.27

The number that you should have obtained in parts (b) and (c) of the previous question is 1.4142135624.
(a) Why is this number the same for 15 and 100 recursions?
(b) Is the number really equal to $\sqrt{2}$?
(c) Can a computer manipulate real numbers?

Solution:
(a) The number is only capable of being computed to a given number of significant figures and to that number of significant figures both answers are the same. Had a greater number of significant figures been available then both answers would not have been the same.
(b) No. The number obtained by the computer is a rational number, whereas $\sqrt{2}$ is an irrational number.
(c) No. A computer can only manipulate rational numbers. Real numbers consist of rational and irrational numbers and a computer cannot handle irrational numbers – only rational approximations to them; this despite the fact that in many languages the word real is used to describe a data type that is in fact only rational.

Exercises

3.26 The assignment statement

$$x := 1 + 1/x$$

where x is a real number can be used recursively within a computer program. As the number of recursions increase the values of x approaches a fixed value ϕ. Use this assignment statement within a recursive for... to... loop to evaluate

(a ϕ

(b) $1/\phi$

(c) $\phi - 1$

after

(i) 5 recursions;

(ii) 50 recursions;

(iii) 100 recursions.

3.27 The number that you should have obtained for ϕ in the previous question is 1.6180339888. This number is an approximation to an irrational number that is called the **golden mean**.

(a) Why is this number the same for 50 and 100 recursions?

(b) Why is the number found only an approximation to the golden mean?

(c) Can a computer find the exact value of the golden mean?

Chapter 4

Counting

OBJECTIVES

When you have completed this chapter you will be able to:

□ distinguish between a list and a set;

□ append two lists together;

□ separate the head of a list from the tail;

□ create lists of length 2 using the Cartesian product of two sets;

□ construct a tree diagram to aid in the evaluation of a Cartesian product;

□ construct lists of arbitrary length;

□ count the number of natural numbers between any two natural numbers;

□ count the number of contiguous sublists obtainable from a given list;

□ use the sigma notation;

□ generate a list by selection from a single set;

□ count the number of lists that can be generated by selection from a single set;

□ count the number of subsets of a given cardinality;

□ express the cardinality of a power set as a sum of combinatorial coefficients;

□ manipulate expressions involving combinatorial coefficients;

□ construct Pascal's triangle;

□ use Pascal's triangle and combinatorial coefficients to expand binomials;

□ demonstrate the cardinality of the power set.

When dealing with abstract ideas such as those that are employed in designing a software system it can be the case that we implicitly assert erroneous assumptions when making statements concerning our view of the world. For example, we say that the area of a circle is πr^2 under the implicit assumption that we can measure the area of a circle when in fact we cannot because it is impossible to quantify the irrational number π.

It is essential that whenever we make an assertion that we can back it up with evidence, and the best evidence of all is the **proof** that the assertion is correct. We not only require to know that an assertion is correct, we also need to be able to prove that it is correct – a need that has been recognized since humankind started thinking about abstraction.

It is said that Pythagoras was the first to discover the notion of an irrational number. By drawing a right-angled triangle whose sides adjacent to the right angle were of unit length he found that the length of the hypotenuse – that side opposite the right angle – was of length $\sqrt{2}$. Now $\sqrt{2}$ is not a rational number. It has no complete decimal representation – or rather it has but it is infinite and unpatterned in extent. Simply to claim the existence of a type of number that cannot be used in any measuring process was insufficient – its existence had to be proved, and this Pythagoras did in his famous eponymous theorem. This event was sufficient to cause a philosophical upheaval in the thinking of the Ancient Greeks by displaying an abstraction that was proven to be contrary to their beliefs in the beauty, harmony and completeness of the rational number system.

In this chapter we lay the foundations of the ground rules for proof by considering perhaps the simplest yet a most powerful method of proof – proof by induction. We shall see the reasoning that is involved and, along the way, learn a few more mathematical facts concerning numbers and the assumed simplest of mathematical procedures – counting.

Lists and list operations

Lists

A list is an ordered collection of items such as a list of numbers, a list of words or a list of tasks. The notation used to denote a list consists of a pair of angular brackets < >, where the individual items that constitute the list are contained within the brackets. For example, the list of the first six non-zero integers in the sequence of **ascending value** is written as

<1, 2, 3, 4, 5, 6>

Notice that this is not the same list as the list

<6, 5, 4, 3, 2, 1>

because here the first six non-zero integers are given in the sequence of **descending value**. The order in which the elements of a list are written is important.

A list is very similar to a set on which an order has been imposed. For example, the set

{p, q, r, s}

is the same set as

{r, p, s, q}

because it contains the same elements; the order in which the elements are written is of no account. For a list, however, the order in which the elements are given is an intrinsic property of the list.

Many times we wish to refer to a set without writing it out in full. In such a situation we use an identifier, just as we did with sets. For example, if we write

L = <p, q, r, s> **and** M = <r, p, s, q>

then we can use L and M to refer to these two different lists.

Operations on lists

Just as two sets can be united to form a third set called the union of the two original sets, so two lists can be joined together to form a third list. However, because the order of the elements of the list forms a property of the list we cannot simply form the union. Instead, we use an operation called APPEND, symbolized by ∥, where the second list is appended to the end of the first list. For example:

L APPEND M = L ∥ M
= <p, q, r, s> ∥ <r, p, s, q>
= <p, q, r, s, r, p, s, q>

Here is displayed another difference between a set and a list. Sets do not have repeated elements but lists can have repeated elements. Notice also that APPEND is not a commutative operation.

Because order is important in a list we need to be able to identify the nature of an element of the list at a given location in the list. To this end the first element of the list is called the **head** of the list and the remaining elements in the list are referred to as the **tail**. These definitions can be used to spawn two further list operations, namely

HEAD L = HEAD <p, q, r, s>
= p

and

$$\begin{aligned} \text{TAIL L} &= \text{TAIL} <p, q, r, s> \\ &= <q, r, s> \end{aligned}$$

Just as we did with sets, to complete the definition of a list we define the empty list as the list with no elements $< >$.

Counting elements in a list

The number of elements in a list is counted by associating each element of the list with its position number in the list. For example, the list

    <alpha, beta, gamma, delta>

has the element alpha in the **first** position and so we associate alpha with the number 1; beta is in the **second** position and so we associate beta with the number 2; gamma is in the **third** position and so we associate gamma with the number 3; delta is in the **fourth** position and so we associate delta with the number 4. Because the highest number in the list of associated numbers is 4 there are four elements in the list.

    This description may seem to be overstating the case, but it is necessary to get over the point that the process of counting the number of elements in a list is achieved by putting the elements in a one-to-one correspondence with the natural numbers, starting with number 1 and increasing by one at a time. The last element in the list is then associated with the highest number in the correspondence, which is the total number of elements in the list.

> The natural numbers are an intellectual construct and must be distinguished from their identifiers, the numerals 1, 2, 3,... However, it is customary to refer to number 1, number 2, etc.

The LEN operator

Counting the number of elements in a list can be done by defining a recursive operation LEN on a list, which yields the length of the list – the number of elements in the list.

    Firstly, we note that because there are no elements in the empty list it has a length zero. Secondly, we note that the length of the TAIL of a list is one less than the length of the original list, so we can write

$$\begin{aligned} \text{LEN L} &= 0 && \text{if } L = <> \\ &= 1 + \text{LEN (TAIL L)} && \text{otherwise} \end{aligned}$$

For example

$$\begin{aligned} \text{LEN} <a, e, i> &= 1 + \text{LEN (TAIL} <a, e, i>) \\ &= 1 + \text{LEN} <e, i> \\ &= 1 + [1 + \text{LEN (TAIL} <e, i>)] \\ &= 1 + [1 + \text{LEN} <i>] \end{aligned}$$

$$= 1 + [1 + (1 + \text{LEN} [\text{TAIL} <i>])]$$
$$= 1 + [1 + (1 + \text{LEN} < >)]$$
$$= 1 + [1 + (1 + 0)]$$
$$= 3$$

Conclusion: A list is a collection of elements that have a definite order. Lists possess a head and a tail and can be appended to each other. The number of elements in a list is evaluated by putting the elements in a one-to-one correspondence with the non-zero natural numbers. They can also be counted by employing the recursive operator LEN.

Worked example 4.1

If L = <s, u, n> and M = <d, a, y> find
(a)   L APPEND M
(b)   TAIL L APPEND TAIL M
(c)   HEAD <<s, u, n>, <d, a, y>> APPEND HEAD <<s, u, n>, <d, a, y>>

Solution:
(a)   L APPEND M = <s, u, n> APPEND <d, a, y>
                          = <s, u, n, d, a, y>
(b)   TAIL L APPEND TAIL M = TAIL <s, u, n> APPEND TAIL <d, a, y>
                                         = <u, n> APPEND <a, y>
                                         = <u, n, a, y>
(c)   HEAD <<s, u, n>, <d, a, y>> APPEND HEAD <<s, u, n>, <d, a, y>>
                = <s, u, n> APPEND <s, u, n>
                = <s, u, n, s, u, n>

Worked example 4.2

Evaluate each of the following:
(a)   HEAD <l, i, s, t>
(b)   TAIL <<l>, <i>, <s>, <t>>
(c)   HEAD <<l, i, s, t>, <o, f>, <l, i, s, t, s>>
(d)   HEAD(TAIL <<l, i, s, t>, <o, f>, <l, i, s, t, s>>)

Solution:
(a)   HEAD <l, i, s, t> = l
(b)   TAIL <<l>, <i>, <s>, <t>> = <<i>, <s>, <t>>
(c)   HEAD <<l, i, s, t>, <o, f>, <l, i, s, t, s>> = <l, i, s, t>
(d)   HEAD(TAIL <<l, i, s, t>, <o, f>, <l, i, s, t, s>>)
             = HEAD(<<o, f>, <l, i, s, t, s>>)
             = <o, f>

Worked example 4.3

Evaluate each of the following:
(a) LEN <l, i, s, t>
(b) LEN(TAIL <l, i, s, t>)
(c) LEN(HEAD[TAIL <<l, i, s, t>, <o, f>, <l, i, s, t, s>>])

Solution:
(a) LEN <l, i, s, t> = 1 + LEN(TAIL <l, i, s, t>)
                       = 1 + LEN(<i, s, t>)
                       = 1 + (1 + LEN[TAIL<i, s, t>])
                       = 1 + (1 + LEN[<s, t>])
                       = 1 + (1 + [1 + LEN{TAIL<s, t>}])
                       = 1 + (1 + [1 + LEN{<t>}])
                       = 1 + (1 + [1 + {1 + LEN(TAIL<t>)}])
                       = 1 + (1 + [1 + {1 + LEN(<>)}])
                       = 1 + (1 + [1 + {1 + 0}])
                       = 1 + (1 + [1 + 1])
                       = 1 + (1 + 2)
                       = 1 + 3
                       = 4
(b) LEN(TAIL <l, i, s, t>) = LEN(<i, s, t>) = 3
(c) LEN(HEAD[TAIL <<l, i, s, t>, <o, f>, <l, i, s, t, s>>])
            = LEN(HEAD[<<o, f>, <l, i, s, t, s>>])
            = LEN(<o, f>)
            = 2

Exercises

4.1 If L = <f, o, l, d> and M = <t, w, o> find
   (a) M APPEND L
   (b) TAIL L APPEND TAIL M
   (c) HEAD<<f, o, l, d>, <t, w, o>> APPEND HEAD <<f, o, l, d>, <t, w, o>>

4.2 Evaluate each of the following:
   (a) HEAD <s, e, t>
   (b) TAIL <<s>, <e>, < t>>
   (c) HEAD <<s, e, t>, <o, f>, <s, e, t, s>>
   (d) HEAD(TAIL <<s, e, t>, <o, f>, <s, e, t, s>>)

4.3 Evaluate each of the following:
   (a) LEN <l, e, n, g, t, h>
   (b) LEN(TAIL <l, e, n, g, t, h>)
   (c) LEN(TAIL[HEAD <<l, i, s, t>, <o, f>, <l, i, s, t, s>>])

The Cartesian product

Lists from sets

Sets of lists can be generated from sets of elements using an operation referred to as the Cartesian product. For example, if A and B are two sets then the Cartesian product A × B of A and B is defined as

$$A \times B = \{<x, y> : x \in A \text{ and } y \in B\}$$

which is a set of lists of length 2 where the first member of each list is an element of A and the second member of each list is an element of B.

For example, if A and B are the two sets

$$A = \{0, 1, 2\} \text{ and } B = \{2, 3\}$$

then the Cartesian product of A and B is given as

$$A \times B = \{<0, 2>, <0, 3>, <1, 2>, <1, 3>, <2, 2>, <2, 3>\}$$

Notice that the Cartesian product is not a commutative operation:

$$A \times B \neq B \times A$$

since

$$B \times A = \{<2, 0>, <2, 1>, <2, 2>, <3, 0>, <3, 1>, <3, 2>\}$$

Such lists of length 2 are also referred to as ordered pairs because they contain pairs of elements in strict order.

Trees

To ensure that all possible ordered pairs have been generated from a Cartesian product we must generate them in an orderly manner. This can be readily achieved with the aid of a diagram. For example, to find the Cartesian product A × B where

$$A = \{a, b, c\} \text{ and } B = \{0, 1\}$$

we construct the branched figure shown in Figure 4.1.

Starting from the top of the figure we see that each of the three lines emanating from the starting point A is labelled with the elements of set A. At the ends of each of these lines at B are two further branch lines, and each line of each pair is labelled with the elements of set B. We can now generate ordered pairs by traversing this structure in an orderly way. For example, starting at A and moving down through the left-hand branch to B we

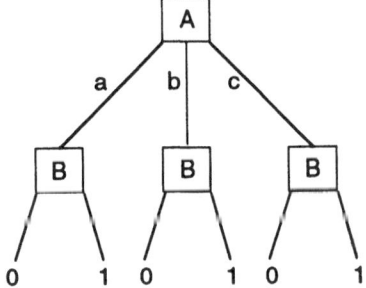

Figure 4.1

traverse the branch labelled a. From here we traverse the left-hand branch labelled 0 to arrive at the bottom of the figure. In this way we have generated the ordered pair <a, 0>. By selecting all possible routes through the figure in an orderly manner we find that all the elements of A × B are generated.

This figure is called a tree diagram because it resembles an upside-down tree. The point at the top labelled A is called the root of the tree and all the lines in the tree are called branches; the particular tree that we have looked at here is called a two-level tree. We shall look at trees in more detail in Chapter 7.

Worked example 4.4

Write down the complete set of lists of length 2 generated by A × B where
(a)   A = {2, 4, 6, 8} and B = {−1, 0, 1}
(b)   A = {a, b, c} and B = {1, 3}
(c)   A = {cow, sheep, pig} and B = {byre, fold, sty}

Solution:
(a)   {<2, −1>, <2, 0>, <2, 1>, <4, −1>, <4, 0>, <4, 1>, <6, −1>, <6, 0>, <6, 1>, <8, −1>, <8, 0>, <8, 1>}
(b)   {<a, 1>, <a, 3>, <b, 1>, <b, 3>, <c, 1>, <c, 3>}
(c)   {<cow, byre>, <sheep, byre>, <pig, byre>, <cow, fold>, <sheep, fold>, <pig, fold>, <cow, sty>, <sheep, sty>, <pig, sty>}

Worked example 4.5

From which two sets can the following sets of lists be generated?
(a)   {<1, a>, <1, b>, <2, a>, <2, b>, <3, a>, <3, b> }
(b)   {<gold, silver>, <silver, tin>, <gold, tin>, <tin, gold>, <tin, silver>, <silver, gold>, <gold, gold>, <tin, tin>, <silver, silver>}

Solution:

(a) A × B where A = {1, 2, 3} and B = {a, b}

(b) A × B where A = {gold, silver, tin} and B = A

Worked example 4.6

Given the sets A = {1, 2, 3, 4} and B = {2, 4, 6, 8} then each of the following sets of lists is a subset of the cross-product A × B. In each case find the condition imposed on the members of each list in A × B that selects that list for membership of the appropriate subset:

(a) {<1, 2>, <1, 4>, <1, 6>, <1, 8>, <2, 2>, <2, 4>, <2, 6>, <2, 8>, <3, 4>, <3, 6>, <3, 8>, <4, 4>, <4, 6>, <4, 8>}

(b) {<1, 2>, <2, 4>, <3, 6>, <4, 8> }

(c) {<3, 2>, <4, 2> }

Solution:

(a) {<1, 2>, <1, 4>, <1, 6>, <1, 8>, <2, 2>, <2, 4>, <2, 6>, <2, 8>, <3, 4>, <3, 6>, <3, 8>, <4, 4>, <4, 6>, <4, 8>}. Here, $x \leq y$ where $x \in$ A and $y \in$ B.

(b) {<1, 2>, <2, 4>, <3, 6>, <4, 8>}. Here, $y = 2x$ where $x \in$ A and $y \in$ B.

(c) {<3, 2>, <4, 2>}. Here, $x > y$ where $x \in$ A and $y \in$ B.

Exercises

4.4 Write down the complete set of lists of length 2 generated by A × B where

(a) A = {3, 5} and B = {−1, −2, −3}

(b) A = {w, x, y} and B = {a, b, c, d}

(c) A = {Mr, Mrs, Ms} and B = {Jones, Robinson, Andrews, McDougal}

4.5 From which two sets can the following sets of lists be generated?

(a) {<−1, p>, <−1, q>, <−1, r>, <0, p>, <0, q>, <0, r>}

(b) {<Bradford, England>, <Vancouver, Canada>, <Sydney, Australia>, <Bradford, Canada>, <Sydney, England>, <Vancouver, Australia>, <Vancouver, England>, <Bradford, Australia>, <Sydney, Canada>}

4.6 Given the sets A = {−1, 0, 1, 2} and B = {0, 1, 3, 4, 5} then each of the following sets of lists is a subset of the cross-product A × B. In each case find the condition imposed on the members of each list in A × B that selects that list for membership of the appropriate subset:

(a) {<2, 1>, <2, 0>, <1, 0>}

(b) {<−1, 1>, <1, 3>, <2, 4>}
(c) {<−1, 1>, <0, 0>}

Lists of length *n* and ordered *n*-tuples

We have seen how lists of length 2 can be generated by application of the Cartesian product between pairs of sets. Repeated Cartesian products can generate lists of any length whatsoever. For example, if

$$A = \{-1, 1\}, B = \{0, 2\} \text{ and } C = \{3, 5\}$$

then by extending the tree diagram to three levels a set of eight lists, each of length 3 or ordered triples can be formed as shown in Figure 4.2.

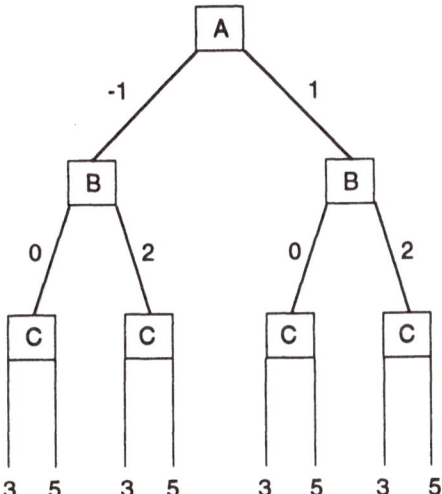

Figure 4.2

$$A \times B \times C = \{<-1, 0, 3>, <-1, 0, 5>, <-1, 2, 3>, <-1, 2, 5>, <1, 0, 3>, <1, 0, 5>, <1, 2, 3>, <1, 2, 5>\}$$

In this manner, by extending the tree structure to *n* levels we can generate lists of length *n*, also known as ordered *n*-tuples.

The use of brackets

The use of brackets in an expression involving the Cartesian product will radically affect the end result. For example, in the expression

$$A \times (B \times C)$$

the brackets indicate that the Cartesian product B × C is to be performed first. This will result in a set of ordered pairs. When the second Cartesian product is evaluated between this set and A the end result is again a set of ordered pairs and not a set of ordered triples. For example, if:

A = {a}, B = {1, 2} and C = {up}

then

A × (B × C) = A × {<1, up>, <2, up>}
            = {<a, <1, up>>, <a, <2, up>>}

Notice that we have not generated ordered triples but ordered pairs where each ordered pair has one of its members as an ordered pair itself. Also

(A × B) × C = {<a, 1>, <a, 2> } × C
            = {<<a, 1>, up>, <<a, 2>, up>}
            ≠ A × (B × C)

which demonstrates that the Cartesian product is not associative.

Worked example 4.7

Given A = {Smith, Hardy, Thomas}, B = {Engineering, Computing, Physics} and C = {Year 1, Year 2} find
(a) A × (B × C)
(b) (A × B) × C
(c) A × B × C

Solution:
(a) B × C = {Engineering, Computing, Physics} × {Year 1, Year 2}
            = {<Engineering, Year 1>, <Engineering, Year 2>,
                <Computing, Year 1>, <Computing, Year 2>,
                <Physics, Year 1>, <Physics, Year 2>}
     A × (B × C) = {<Smith, <Engineering, Year 1>>, <Smith,
                <Engineering, Year 2>>, <Smith, <Computing,
                Year 1>>, <Smith, <Computing, Year 2>>,
                <Smith, <Physics, Year 1>>, <Smith, <Physics,
                Year 2>>, <Hardy, <Engineering, Year 1>>,
                <Hardy, <Engineering, Year 2>>, <Hardy,
                <Computing, Year 1>>, <Hardy, <Computing,
                Year 2>>, <Hardy, <Physics, Year 1>>, <Hardy,
                <Physics, Year 2>>, <Thomas, <Engineering,
                Year 1>>, <Thomas, <Engineering, Year 2>>,

                    <Thomas, <Computing, Year 1>>, <Thomas, <Computing, Year 2>>, <Thomas, <Physics, Year 1>>, <Thomas, <Physics, Year 2>>}

(b) $(A \times B) \times C$ = {<<Smith, Engineering>, Year 1>, <<Smith, Engineering>, Year 2>, <<Smith, Computing>, Year 1>, <<Smith, Computing>, Year 2>, <<Smith, Physics>, Year 1>, <<Smith, Physics>, Year 2>, <<Hardy, Engineering>, Year 1>, <<Hardy, Engineering>, Year 2>, <<Hardy, Computing>, Year 1>, <<Hardy, Computing>, Year 2>, <<Hardy, Physics>, Year 1>, <<Hardy, Physics>, Year 2>, <<Thomas, Engineering>, Year 1>, <<Thomas, Engineering>, Year 2>, <<Thomas, Computing>, Year 1>, <<Thomas, Computing>, Year 2>, <<Thomas, Physics>, Year 1>, <<Thomas, Physics>, Year 2>}

(c) $A \times B \times C$ = {<Smith, Engineering, Year 1>, <Smith, Engineering, Year 2>, <Smith, Computing, Year 1>, <Smith, Computing, Year 2>, <Smith, Physics, Year 1>, <Smith, Physics, Year 2>, <Hardy, Engineering, Year 1>, <Hardy, Engineering, Year 2>, <Hardy, Computing, Year 1>, <Hardy, Computing, Year 2>, <Hardy, Physics, Year 1>, <Hardy, Physics, Year 2>, <Thomas, Engineering, Year 1>, <Thomas, Engineering, Year 2>, <Thomas, Computing, Year 1>, <Thomas, Computing, Year 2>, <Thomas, Physics, Year 1> <Thomas, Physics, Year 2>}

Worked example 4.8

List the sets and their product form that are required to create the following sets of lists:

(a) {<ham, apple, coffee>, <beef, orange, tea>, <ham, apple, juice>, <ham, orange, coffee>, <beef, apple, tea>, <ham, orange, juice>, <ham, apple, tea>, <beef, orange, coffee>, <beef, orange, juice>, <ham, orange, tea>, <beef, apple, coffee>, <beef, apple, juice>}

(b) {<<ham, apple>, coffee>, <<beef, orange>, tea>, <<ham, apple>, juice>, <<ham, orange>, coffee>, <<beef, apple>, tea>, <<ham, orange>, juice>, <<ham, apple>, tea>, <<beef, orange>, coffee>, <<beef, apple>, juice>, <<ham, orange>, tea>, <<beef, apple>, coffee>, <<beef, orange>, juice>}

Solution: The first elements of each list are selected from the set A = {ham, beef}.The second elements are from B = {orange, apple} and the third elements are from C = {coffee, tea, juice}. Consequently, from the location of the lists' brackets, we see that

(a) A × B × C

(b) (A × B) × C

Exercises

4.7 Given A = {Saturday, Sunday}, B = {May, June, July} and C = {cricket, swimming, golf} find

(a) A × (B × C)

(b) (A × B) × C

(c) A × B × C

4.8 List the sets and their product form that are required to create the following sets of lists:

(a) {<Ford, three-door, blue>, <Vauxhall, four-door, red>, <Ford, three-door, white>, <Ford, four-door, blue>, <Vauxhall, three-door, red>, <Ford, four-door, white>, <Ford, three-door, red>, <Vauxhall, four-door, blue>, <Vauxhall, four-door, white>, <Ford, four-door, red>, <Vauxhall, three-door, blue>, <Vauxhall, three-door, white>}

(b) {<<Ford, three-door>, blue>, <<Vauxhall, four-door>, red>, <<Ford, three-door>, white>, <<Ford, four-door>, blue>, <<Vauxhall, three-door>, red>, <<Ford, four-door>, white>, <<Ford, three-door>, red>, <<Vauxhall, four-door>, blue>, <<Vauxhall, three-door>, white>, <<Ford, four-door>, red>, <<Vauxhall, three-door>, blue>, <<Vauxhall, four-door>, white>}

Counting numbers and contiguous sublists

Counting

Imagine you are sitting on an open farm gate counting sheep as they pass through. For each sheep that walks past you look at it and count: one, two, three,... What you are doing is mentally associating a unique number with each sheep; you are placing the sheep in a **one-to-one correspondence with the natural numbers**. The highest number you count to represents the total number of sheep that have walked past.

To count the number of colours in the rainbow we form a one-to-one correspondence between the colours and the natural numbers:

Red   → 1
Orange   → 2

Yellow → 3
Green → 4
Blue → 5
Indigo → 6
Violet → 7

From this we see that there are seven colours in the rainbow. This is how we count.

Numbers of numbers

We have already seen how numbers can be used to count by putting the items we wish to count into a **one-to-one correspondence with the natural numbers**. The same can be done to permit the counting of numbers themselves. For example, to find the number of numbers between 10 and 15 inclusive we create the **correspondence**:

10 → 1
11 → 2
12 → 3
13 → 4
14 → 5
15 → 6

from which we see that there are six numbers between 10 and 15 inclusive. Notice that:

$$6 = 15 - 10 + 1$$

That is, the number of numbers is the difference between the first and last numbers plus 1.

Is it possible to use the form of this equation to give the number of natural numbers between and inclusive of any pair of natural numbers? We are tempted to say yes and that the number of natural numbers between the natural numbers n and r inclusive is $N[n, r]$ where $N[n, r]$ is given by the formula

$$N[n, r] = n - r + 1 \text{ and where } 0 \leq r \leq n$$

We can test this formula for any pair of natural numbers and find that it gives the correct answer. For example

$$N[7, 3] = 7 - 3 + 1 = 5 \text{ and } N[9, 0] = 9 - 0 + 1 = 10$$

However, simply substituting numbers into the formula does not prove that the formula is correct: it only demonstrates that it is correct for those numbers we choose to test it with. To prove that it is correct for any pair of numbers n and r we must take a different approach, known as proof by induction.

The ability to prove our statements correct is crucial to proper software construction. Proof by induction is our introduction to the whole idea of proof and it is essential that you understand the reasoning employed here.

Proof by induction

To prove that the formula is correct we first make the assumption that the formula **is** correct, namely that the number of natural numbers between the natural numbers r and n is $N[n, r]$ where

$N[n, r] = n - r + 1$ and where r is some fixed number such that $0 \leq r \leq n$

Next, under the assumption that the formula is correct we find that the number of natural numbers between the natural numbers r and $(n + 1)$ is one more and is given as

$$
\begin{aligned}
N[n + 1, r] &= N[n, r] + 1 \\
&= n - r + 1 + 1 \\
&= (n + 1) - r + 1
\end{aligned}
$$

The form of this result, namely

$$N[n + 1, r] = (n + 1) - r + 1$$

shows that the form of the formula is maintained with n being replaced by $n + 1$ on both sides of the equation.

Consequently, we have proved that if the formula is true for r and n then it must be true for r and $n + 1$.

Finally, we see that when $n = r$

$$
\begin{aligned}
N[r, r] &= r - r + 1 \\
&= 1
\end{aligned}
$$

which is clearly true – the number of natural numbers between the natural number r and itself inclusive is 1.

In conclusion, we have shown that the formula is

TRUE when $n = r$ and by our proof it must also be TRUE when $n = r + 1$

Hence, since it is

TRUE when $n = r + 1$ it must be TRUE when $n = r + 2$

Hence, since it is

TRUE when $n = r + 2$ it must be TRUE when $n = r + 3$

And so on. Which proves that the formula is true for any value of $n \geq r$ where r is fixed (notice that $n \geq r \equiv r \leq n$). Furthermore, despite the fact that we stated that r was a fixed natural number it was never specified as to which natural number it was. Therefore it can be any natural number less than or equal to n. Consequently, the formula is true for any value of r and any value of n provided $0 \leq r \leq n$.

This method of mathematical proof is known as proof by induction because the proof is induced from the form of the formula itself. It is a very powerful method of proof and we shall see more instances of it as we progress.

Worked example 4.9

Prove each of the following formulae by induction:
(a) $3 + 6 + 9 + \dots + 3n = [3/2]n(n + 1)$
(b) $1/(1 \times 2) + 1/(2 \times 3) + 1/(3 \times 4) + \dots + 1/(n[n + 1]) = n/(n + 1)$

Solution:
(a) Assume that
$$3 + 6 + 9 + \dots + 3n = [3/2]n(n + 1)$$
Then
$$3 + 6 + 9 + \dots + 3n + 3(n + 1) = [3/2]n(n + 1) + 3(n + 1)$$
$$\text{adding } 3(n + 1)$$
$$= 3n^2/2 + 3n/2 + 3n + 3$$
$$\text{multiplying out}$$
$$= 3n^2/2 + 9n/2 + 3$$
$$\text{simplifying}$$
$$= [1/2](3n^2 + 9n + 6)$$
$$\text{factorizing out } 1/2$$
$$= [1/2](3n + 3)(n + 2)$$
$$\text{factorizing}$$
$$= [3/2](n + 1)([n + 1] + 1)$$
$$\text{rewriting}$$
This is the same form as the original with n replaced by $n + 1$. Consequently, if the formula is true for n then it is true for $n + 1$.

When $n - 1$

Left-hand side $= 3 \times 1 = 3$ and right-hand side $= [3/2]1(1 + 1) = 3$

Hence the formula is true when $n = 1$, therefore it is true when $n = 1 + 1 = 2$, therefore it is true when $n = 2 + 1 = 3$, and so on.
(b) Assume that

$$1/(1 \times 2) + 1/(2 \times 3) + 1/(3 \times 4) + \dots + 1/(n[n + 1]) = n/(n + 1)$$

Then

$$1/(1 \times 2) + 1/(2 \times 3) + 1/(3 \times 4) + \dots + 1/(n[n + 1]) + 1/[(n + 1)(n + 2)]$$
$$= \ n/(n + 1) + 1/(n + 1)(n + 2) \qquad \text{adding } 1/[(n + 1)(n + 2)]$$

$$= [1/(n+1)](n + 1/(n+2))) \qquad \text{factorizing out } 1/(n+1)$$
$$= [1/(n+1)][n(n+2) + 1]/(n+2) \quad \text{adding } n \text{ to } 1/(n+2)$$
$$= \{1/[(n+1)(n+2)]\}[n(n+2) + 1] \text{ factorizing out } 1/(n+2)$$
$$= \{1/[(n+1)(n+2)]\}[n^2 + 2n + 1] \text{ multiplying out } n(n+2) + 1$$
$$= \{1/[(n+1)(n+2)]\}[n+1]^2 \qquad \text{rewriting } n(n+2) + 1 \text{ as } [n+1]^2$$
$$= (n+1)/(n+2) \qquad\qquad\qquad \text{cancelling } n+1 \text{ top and bottom}$$
$$= (n+1)/([n+1] + 1) \qquad\qquad \text{rewriting}$$

This is the same form as the original with n replaced by $n + 1$. Consequently, if the formula is true for n then it is true for $n + 1$.

When $n = 1$

Left-hand side $= 1/(1 \times 2) = 1/2$ and right-hand side $= 1/(1 + 1) = 1/2$

Hence the formula is true when $n = 1$, therefore it is true when $n = 1 + 1 = 2$, therefore it is true when $n = 2 + 1 = 3$, and so on.

Exercise

4.9 Prove each of the following formulae by induction:
   (a) $2 + 4 + 8 + \ldots + 2^n = 2(2^n - 1)$
   (b) $1^2 + 2^2 + 3^2 + \ldots + n^2 = [n/6](n + 1)(2n + 1)$

The word contiguous is derived from the Latin *tangere* which means to touch.

Numbers of contiguous sublists

Given the list L:

   L = <a, b, c, d, e, f>

then the lists M and N where

   M = <b, c, d> and N = <a, c, d>

are called sublists of list L; a **sublist** of a list is any list consisting of members of the original list where the order is preserved. For example

   P = <e, d, f>

is not a sublist of L because the order of the members is not the same as the order of the same members in L.

   A **contiguous** sublist is a sublist whose adjacent members are also adjacent members of the original list. Consequently, M is a contiguous sublist of list L because all adjacent members of M are also adjacent members of L. However, sublist N is not contiguous because the adjacent members a and c of N are not adjacent members of L.

   Notice that there are only four contiguous sublists of length 3 possible from L. Selecting the contiguous sublists in a strict order by taking the first sublist from the first three members of L and the second sublist by moving down a member and so on we find the contiguous sublists of length 3 are

<a, b, c>, <b, c, d>, <c, d, e>, <d, e, f>

That is, there are

$$4 = 6 - 3 + 1$$

contiguous sublists of length 3 possible from a list of length 6. We have been here before. Given a list of length n then, the number of contiguous sublists of length r that can be created from it is given as $N[n, r]$ where

$$N[n, r] = n - r + 1$$

The proof of this statement is as follows. Let L be a list of length n. The first contiguous sublist of length r will be composed of the first r members of L, leaving $n - r$ members of L remaining. The next contiguous sublist will be composed of members of L taken by moving down the list L by one member. This can be done $n - r$ times so that the total number of contiguous sublists is the first one plus $n - r$ more. That is

$$N[n, r] = n - r + 1$$

Repetitive counting

The addition of natural numbers is well accepted. For example, adding six 1's together produces the result 6:

$$1 + 1 + 1 + 1 + 1 + 1 = 6$$

This notation for repetitive addition soon becomes too cumbersome to write down, and to be more compact we can use what is called the **sigma notation** (Σ-notation)

$$\sum_{r=1}^{6} 1 = 1 + 1 + 1 + 1 + 1 + 1 = 6$$

Here the symbol Σ is an upper-case Greek letter s to indicate the sum of 1's. The letter r beneath the Σ is a counter that ranges from $r = 1$ to $r = 6$ – the 6 above the Σ indicates the range of the count, starting with the first 1 (when $r = 1$) and ending with the sixth 1 (when $r = 6$).

Repetitive addition is used to define the arithmetic operation of multiplication of two natural numbers. For example, if we were to add five 3's together then we would arrive at the result 15:

$3 + 3 + 3 + 3 + 3 = 15$

The addition of five 3's is defined as the product of 3×5. That is

$3 + 3 + 3 + 3 + 3 = 3 \times 5 = 15$

In this case we can employ the Σ notation as

$$\sum_{r=1}^{5} 3 = 3 + 3 + 3 + 3 + 3 = 3 \times 5 = 15$$

Here, we are summing the number 3 five times as the counter number r ranges from 1 to 5.

In general, we can say that:

$$\sum_{r=1}^{n} k = kn$$

where k is a given constant number.

Notice that if $k = 1$ then the sum is simply n. This formula can be proved to be correct by induction.

A more interesting case arises when the quantity being added is related to the counting number r. For example, the sum

$1 + 2 + 3 + 4 = 10$

can be represented as

$$\sum_{r=1}^{4} r = 10$$

Here the quantity being added is equal to the count number and the sum represents the sum of the first four non-zero natural numbers.

What is of specific interest now is: what is the value of the first n non-zero natural numbers?

$$1 + 2 + 3 + \ldots + n = \sum_{r=1}^{n} r$$

The answer to this question is

$$\sum_{r=1}^{n} r = (1/2)n(n + 1)$$

For example

$$\sum_{r=1}^{4} r = (1/2)4(4 + 1) = 2 \times 5 = 10$$

The next question is: How did we know that?

The formula for the sum of the first n non-zero integers can be derived using a knowledge of arithmetic sequences and series. Here, however, we shall not derive the formula, instead we shall prove it to be correct.

'How can we prove something correct without deriving it?' you may ask, and the answer to that question lies in the power of mathematical proof – again, proof by induction.

The first step is to assume that the formula is correct, namely

$$\sum_{r=1}^{n} r = (1/2)n(n + 1)$$

The second step is to find the formula for the sum of the first $n + 1$ non-zero natural numbers:

$$\sum_{r=1}^{n+1} r = \sum_{r=1}^{n} r + (n + 1)$$

$$= (1/2)n(n + 1) + (n + 1)$$
$$= (n + 1)([1/2]n + 1) \qquad \text{factorizing out the } (n + 1)$$
$$= (n + 1)([1/2]n + [1/2]2) \qquad \text{recognizing that } 1 = [1/2]2$$
$$= [1/2](n + 1)(n + 2) \qquad \text{factorizing out the } [1/2]$$
$$= [1/2](n + 1)([n + 1] + 1) \qquad \text{rewriting } n + 2 \text{ as } [n + 1] + 1$$

This is the identical formula to the sum of the first n non-zero integers with n replaced by $n + 1$.

The conclusion that can now be drawn is that

IF the formula is correct for the first n non-zero natural numbers
THEN it is correct for the first $n + 1$ non-zero natural numbers
BECAUSE the form is the same in both cases

The third step is to show that the formula is correct when $n = 1$:

$$\sum_{r=1}^{1} r = [1/2]1(1 + 1) = 1$$

which demonstrates that, indeed, it is correct when $n = 1$.

Finally we conclude that because it is

TRUE for $n = 1$ we have proved it to be TRUE for $n = 1 + 1 = 2$

and because it is

TRUE for $n = 2$ we have proved it to be TRUE for $n = 2 + 1 = 3$

and because it is

TRUE for $n = 3$ we have proved it to be TRUE for $n = 3 + 1 = 4$

and so on forever. In other words, we have proved it to be the correct formula for any non-zero natural number, therefore it can be said to be correct for all non-zero natural numbers n.

Total number of sublists

We have already seen that the number of contiguous sublists of length r that can be created from a list of length n is

$$n - r + 1$$

We now wish to know how many contiguous sublists in total can be created from the list of length n. For example, there are

$n - 1 + 1$ contiguous sublists of length 1
$n - 2 + 1$ contiguous sublists of length 2
$n - 3 + 1$ contiguous sublists of length 3, etc.

In total there are

$$\sum_{r=1}^{n} (n - r + 1)$$

contiguous sublists in total.

Now

$$\sum_{r=1}^{n} (n - r + 1) = \sum_{r=1}^{n} n - \sum_{r=1}^{n} r + \sum_{r=1}^{n} 1$$
$$= n^2 - [1/2]n(n + 1) + n$$

$$= n^2 - [1/2]n^2 - [1/2]n + n$$

multiplying out $(n + 1)$ by $[1/2]n$

$$= [1/2]n^2 + [1/2]n$$
$$= [1/2]n(n + 1) \qquad \text{factorizing } [1/2]n$$

Notice that this does not include the empty list – the list of length 0. Consequently, the total number of contiguous sublists that can be created from a list of length n is

$$[1/2]n(n + 1) + 1$$

Worked example 4.10

Write down the first four terms of each of the following sums:

(a) $\displaystyle\sum_{r=0}^{n} 6^r$

(b) $\displaystyle\sum_{r=1}^{n} (5 - 4r)$

(c) $\displaystyle\sum_{r=1}^{n} r(r + 1)$

Solution:
(a) $6^0 = 1, 6^1 = 6, 6^2 = 36, 6^3 = 216$
(b) $(5 - 4) = 1, (5 - 4 \times 2) = -3, (5 - 4 \times 3) = -7, (5 - 4 \times 4) = -11$
(c) $1(1 + 1) = 2, 2(2 + 1) = 6, 3(3 + 1) = 12, 4(4 + 1) = 20$

Worked example 4.11

Find the general term of each of the following sequences and write the sum of the first n terms in Σ-notation:
(a) $1, 3, 5, 7, 9,\ldots$
(b) $1/3, 1/9, 1/27, 1/81,\ldots$
(c) $1, 1/2, 1/3, 1/4,\ldots$

Solution:
(a) $2r + 1$ starting from $r = 0$

$$\sum_{r=0}^{n-1} (2r + 1)$$

(b) $1/3^r$ starting from $r = 1$

$$\sum_{r=1}^{n} 1/3^r$$

(c) $1/r$ starting from $r = 1$

$$\sum_{r=1}^{n} 1/r$$

Worked example 4.12

Prove that

$$\sum_{r=1}^{n} 1/r^3 = [n^2/4](n + 1)^2$$

Solution: Proof is by induction. Assume

$$\sum_{r=1}^{n} r^3 = [n^2/4](n + 1)^2$$

Then

$$\sum_{r=1}^{n+1} r^3 = \sum_{r=1}^{n} r^3 + (n + 1)^3$$

$$= [n^2/4](n + 1)^2 + (n + 1)^3$$
$$= (n + 1)^2([n^2/4] + (n + 1))$$
$$= [(n + 1)^2/4](n^2 + 4n + 4)$$
$$= [(n + 1)^2/4](n + 2)^2$$
$$= [(n + 1)^2/4]([n + 1] + 1)^2$$

This is the same form as the original with n replaced by $n + 1$. Consequently, if the formula is true for n then it is true for $n + 1$.

When $n = 1$

Left-hand side $= 1^3 = 1$ and right-hand side $= [1/4](1 + 1)^2 = 1$

Hence the formula is true when $n = 1$, therefore it is true when $n = 1 + 1 = 2$, therefore it is true when $n = 2 + 1 = 3$, and so on.

Exercises

4.10 Write down the first four terms of each of the following sums:

(a) $\sum_{r=0}^{n} [1/5]^r$

(b) $\sum\limits_{r=0}^{n} (2r + 3)$

(c) $\sum\limits_{r=1}^{n} \{1/r - 1/(r + 1)\}$

4.11 Find the general term of each of the following sequences and write the sum of the first n terms in Σ-notation:
  (a) 4, 8, 12, 16,...
  (b) 7, 49, 343, 2401,...
  (c) 1, -1, 1, -1,...

4.12 Prove that

$$\sum\limits_{r=1}^{n} r(r + 1) = \lfloor n/3 \rfloor (n + 1)(n + 2)$$

Permutations and combinations

Lists from a single set

We have seen how lists can be generated from sets by using the Cartesian product. This is not the only way to create a list. It is also possible to create a list from a single set purely by selection. For example, given the set

  {Monday, Wednesday, Friday, Sunday}

we can generate the lists

  <Monday, Friday, Sunday> and <Sunday, Monday, Wednesday, Friday>

simply by selecting elements from the set. Indeed, we could also form the list

  <Friday, Monday, Friday>

by repetitive selection. Our next question is: How many lists can be generated from a given set of elements? In the first stage towards answering this question we shall consider how many lists of different members can be created from a given collection of elements. For example, how many lists of length 3 can be created from the set of three elements using each element once only in a given list?

  {apple, orange, pear}

The following are all the possible lists:

<apple, orange, pear>, <orange, apple, pear>, <pear, apple, orange>, <apple, pear, orange>, <orange, pear, apple>, <pear, orange, apple>

Consider how elements were selected to create these lists. There are three ways of selecting the first element, hence the three columns of lists. Having selected the first element there are just two elements left from which to choose the remaining elements, hence the two rows of lists. This means that there are

$$3 \times 2$$

different ways of selecting the first two elements. Having selected the first two elements there is no choice for the third element as there is only one element left to choose from. Hence there are

$$3 \times 2 \times 1 = 6$$

different lists that can be constructed from the three elements. This product has a special name. It is called 3 **factorial** and is denoted by the notation 3!, that is

$$3! = 3 \times 2 \times 1$$

In general

$$n! = n \times (n - 1) \times (n - 2) \times \ldots \times 3 \times 2 \times 1$$

where $n \in \mathcal{N}$. Consequently, there are $n!$ different lists that can be formed from n different elements using each element once only in any given list.

Now we are ready to find out how many lists of length r can be created from a given set of n elements where $r \leq n$ and where, in any given list, there are no repeated members. For example, how many lists of length 3 can be created from the set

$$\{a, b, c, d, e, f, g\}$$

The answer to this question again lies in the actual method employed in creating a list. To create a list a first element has to be chosen, and this can be selected from all seven characters in the list, that is there are seven different ways of selecting the first character of the sublist.

Having chosen the first character for the sublist there are now only six characters to choose from in the original list. Hence for each selection of the first character of the sublist there are six different ways of selecting the second character. That is there are $7 \times 6 = 42$ different ways of selecting the first two characters of the three-character sublist. Having selected the first two characters there are now only five characters left in the original list. So for each of the 42 ways of selecting the first two characters there are five different ways of selecting the third. In other words, there are

$7 \times 6 \times 5 = 210$

different ways of selecting the three characters to form the required sublist. That is there are

$7 \times 6 \times 5 = 210$

different three-character sublists possible from an original seven-character set. Now

$$7 \times 6 \times 5 = [7 \times 6 \times 5 \times 4 \times 3 \times 2 \times 1]/[4 \times 3 \times 2 \times 1]$$
$$= 7!/4!$$
$$= 7!/(7-3)!$$

In general if the original set has n elements then there are

$n!/(n-r)!$

ways of selecting an r character list where $r \leq n$. This quantity is denoted by ^nP_r and is called the number of permutations of r items from n distinct items. Notice that we must cater for the case when $r = n$, that is when $n - r = 0$. We do this by defining

$0! = 1$

The factorial is a special case of a more general entity known as the gamma function $\Gamma(x)$, which is a recursive function defined in terms of a semi-infinite integral. When the argument x is a natural number the gamma function coincides with the factorial: $\Gamma(n + 1) = n!$ For this reason it is consistent to define $0! = 1$.

Worked example 4.13

Find the value of
(a) 4!
(b) 6!/3!
(c) 8!/(8 − 2)!

Solution:
(a) $4! = 4 \times 3 \times 2 \times 1 = 24$
(b) $6!/3! = (6 \times 5 \times 4 \times 3 \times 2 \times 1)/(3 \times 2 \times 1) = 6 \times 5 \times 4 = 120$
(c) $8!/(8 - 2)! = 8!/6! = 8 \times 7 = 56$

Worked example 4.14

Find the value of:

(a) $n!/(n-1)!$
(b) $(n+1)!/(n-1)!$

Solution:
(a) $n!/(n-1)! = [n \times (n-1) \times (n-2) \times ... \times 3 \times 2 \times 1]/[(n-1)$
$\times (n-2) \times ... \times 3 \times 2 \times 1]$
$= n$
(b) $(n+1)!/(n-1)! = (n+1)(n)[(n-1)!]/(n-1)!$
$= (n+1)(n)$
$= n^2 + n$

Worked example 4.15

How many two-character lists can be created by selection from

  $\{a, b, c, d, e, f, g, h, i, j\}$

where there is no repetition of members in any list.

Solution: The first character can be selected in any one of 10 ways and the second can be selected from among any one of the nine remaining characters. Therefore there are:

  $10 \times 9 = 90$

different two-character lists that can be formed from the 10-character set where no character is repeated in any given list. This is in agreement with the number $n!/(n-r)!$ where $n = 10$ and $r = 2$.

Worked example 4.16

How many two-character lists can be created by selection from

  $\{1, 2, 3, ..., n\}$

where there is no repetition of members in any list.

Solution: The number of two-character lists (without repetition) that can be created from a set of n characters is $n!/(n-2)! = n(n-1)$.

Exercises

4.13 Find the value of
  (a) 5!

    (b) $9!/4!$

    (c) $7!/(7 - 4)!$

4.14 Find the value of:

    (a) $(n + 1)!/n!$

    (b) $(n - 1)!/(n + 1)!$

4.15 How many three-number lists can be created by selection from

$$\{1, 2, 3, 4, 5, 6\}$$

where there is no repetition of members in any list?

4.16 How many four-number lists can be created by selection from

$$\{1, 2, 3,\dots, n - 1\}$$

where there is no repetition of members in any list?

Counting subsets

The number of subsets of a given cardinality that can be formed from a set differs from the number of lists that can be formed from the set because, in a set, the order of the appearance of the elements does not matter. For example, the number of subsets of cardinality 3 that can be formed from the set

$$\{a, b, c, d, e, f, g\}$$

is less than the corresponding number of three-character lists that can be formed. Take, for instance, the subset

$$\{a, b, c\}$$

There are $3 \times 2 \times 1 = 3!$ different ways of arranging the elements in this subset, each arrangement producing the identical subset. Consequently, for each three-character selection there are $3!$ different arrangements that produce the same subset. As there are a total of

$$7!/(7 - 3)!$$

subsets where each subset is repeated $3!$ times then there are

$$7!/[(7 - 3)!3!]$$

different subsets. This number is referred to as the number of different combinations of three items from seven different items and is denoted by $^{7}C_{3}$. In general, there are

$$^{n}C_{r} = n!/[(n - r)!r!]$$

The so-called permutations used by football pools promoters are, in fact, combinations.

different combinations of r items out of n different items.

The power set
The power set of a set S is the set of all possible subsets of S. Clearly, if S has n elements then there are

$^nC_0$ subsets with 0 elements
$^nC_1$ subsets with 1 element
$^nC_2$ subsets with 2 elements
...
^nC_r subsets with r elements
...
^nC_n subsets with n elements

So that, in total, there are

$$\sum_{r=0}^{n} {}^nC_r$$

subsets that can be generated from a set of n elements. This is the cardinality of the power set.

Worked example 4.17

Evaluate
(a) $7!/[5!2!]$
(b) $8!/[(8-3)!3!]$
(c) $^5C_2$
(d) $\sum_{r=0}^{5} {}^5C_r$

Solution:
(a) $7!/[5!2!] = (7!/5!)(1/2!) = (7 \times 6)/(2 \times 1) = 21$
(b) $8!/[(8-3)!3!] = (8!/5!)(1/3!) = (8 \times 7 \times 6)/(3 \times 2 \times 1) = 56$
(c) $^5C_2 = 5!/[(5-2)!2!] = (5!/3!)(1/2!) = (5 \times 4)/(2 \times 1) = 10$

(d) $\sum_{r=0}^{5} {}^5C_r = {}^5C_0 + {}^5C_r + {}^5C_2 + {}^5C_3 + {}^5C_4 + {}^5C_5$

$= 5!/[(5-0)!0!] + 5!/[(5-1)!1!] + 5!/[(5-2)!2!] + 5!/[(5-3)!3!]$
$\qquad + 5!/[(5-4)!4!] + 5!/[(5-5)!5!]$
$= 5!/[5!0!] + 5!/[4!1!] + 5!/[3!2!] + 5!/[2!3!] + 5!/[1!4!]$
$\qquad + 5!/[0!5!]$
$= 1 + 5 + 10 + 10 + 5 + 1$
$= 32$

Worked example 4.18

How many three-element subsets are there of the following set?

A = {one, two, three, four, five, six, seven}

Solution: The number of subsets of cardinality 3 that can be formed from a set of cardinality 7 is.

$$^7C_3 = 7!/[(7 - 3)!3!]$$
$$= (7 \times 6 \times 5)/(3 \times 2 \times 1)$$
$$= 35$$

Worked example 4.19

What is the cardinality of the power set of the following set?

A = {Ford, Volvo, Toyota, Honda, Mitsubishi, Chevrolet, Renault}

Solution: The cardinality of the power set of a set with cardinality 7 is

$$\sum_{r=0}^{7} {}^7C_r = 128 = 2^7$$

Exercises

4.17 Evaluate
  (a) $9!/[6!3!]$
  (b) $10!/[(10 - 2)!2!]$
  (c) $^4C_3$
  (d) $\sum_{r=0}^{4} {}^4C_r$

4.18 How many four-element subsets are there of the following set?

A = {red, orange, yellow, green, blue, indigo, violet}

4.19 What is the cardinality of the power set of the following set?

A = {a, e, i, o, u}

Lists containing identical members
Unlike sets, lists can have repeated members and such lists can be generated from sets by using repetitive selection. For example, given the set

{a, b}

then

<a, a>, <a, b>, <b, a> and <b, b>

is the complete collection of two-member lists that can be generated from the set by using repetitive selection and

<a, a, a>, <a, a, b>, <a, b, a>, <b, a, a,>, <a, b, b>, <b, a, b>, <b, b, a>, <b, b, b>

is the complete collection of three-member lists that can be generated from the set by using repetitive selection. This now raises the question: Given an n-member list composed of two different types of member a and b how many lists are there containing r a's?

To answer this question consider the process of selecting lists from a two-element set by using repetitive selection. Because there are n members to the list there would be, if all the members were different, $n!$ different lists. However, because there are only two different types of member, namely a and b, many of these $n!$ arrangements are identical. In any given arrangement of a's and b's the a's can be rearranged among themselves in $r!$ ways and the b's can be rearranged among themselves in $(n - r)!$ ways to leave the list unchanged. Consequently there are just

$$n!/\{[n - r]!r!\} = {}^nC_r$$

different arrangements of r a's and $n - r$ b's. From this result it can be seen that the total number of lists of length n that can be generated from a set containing just two elements is

${}^nC_0$ lists with 0 a's
${}^nC_1$ lists with 1 a
${}^nC_2$ lists with 2 a's
...
${}^nC_r$ lists with r a's
...
${}^nC_n$ lists with n a's

In total there are

$$\sum_{r=0}^{n} {}^nC_r$$

lists of length n that can be generated from a set containing just two elements. Notice that this is the same as the cardinality of the power set of a

set composed of n elements – the total number of subsets that can be generated from a given set of n elements.

Worked example 4.20

Write down all the lists of length 4 that can be created by repetitive selection from the set

$\{0, 1\}$

and show that the number of lists is equal to

$$\sum_{r=0}^{4} {}^{4}C_r$$

Solution:

<0, 0, 0, 0>, <0, 0, 0, 1>, <0, 0, 1, 0>, <0, 0, 1, 1>, <0, 1, 0, 0>, <0, 1, 0,
1>, <0, 1, 1, 0>, <0, 1, 1, 1>, <1, 0, 0, 0>, <1, 0, 0, 1>, <1, 0, 1, 0>, <1, 0,
1, 1>, <1, 1, 0, 0>, <1, 1, 0, 1>, <1, 1, 1, 0>, <1, 1, 1, 1>

16 lists.

$$\sum_{r=0}^{4} {}^{4}C_r = {}^{4}C_0 + {}^{4}C_1 + {}^{4}C_2 + {}^{4}C_3 + {}^{4}C_4$$
$$= 1 + 4 + 6 + 4 + 1$$
$$= 16$$

Worked example 4.21

How many binary numbers are there between and including 0000 and 1111?

Solution: Each place in the binary number can be filled in two ways, therefore there are $2 \times 2 \times 2 \times 2 = 2^4 = 16$ different ways the binary number can
be formed.

Worked example 4.22

How many hexadecimal numbers are there between and including 00 and
FF where the numbers are generated by repetitive selection from the following set?

$\{0, 1, 2, 3, 4, 5, 6, 7, 8, 9, A, B, C, D, E, F\}$

Solution: Each place in the hexadecimal number can be filled in 16 ways, therefore there are $16 \times 16 = 16^2 = 256$ different ways the hexadecimal number can be formed.

Exercises

4.20 Write down all the lists of length 3 that can be created by repetitive selection from the set

$\{0, 1\}$

and show that the number of lists is equal to

$$\sum_{r=0}^{3} {}^3C_r$$

4.21 How many binary numbers are there between and including 000 and 111?

4.22 How many hexadecimal numbers are there between and including 000 and FFF where the numbers are generated by repetitive selection from the following set?

$\{0, 1, 2, 3, 4, 5, 6, 7, 8, 9, A, B, C, D, E, F\}$

Combinatorics

Properties of combinations

The number of combinations of r objects out of n different objects, namely

$${}^nC_r = n!/\{[n-r]!r!\}$$

has a number of properties. For example

1. ${}^nC_r = {}^nC_{n-r}$
2. ${}^nC_0 = {}^nC_n = 1$
3. ${}^{n+1}C_1 - {}^nC_1 = 1$
4. ${}^nC_r + {}^nC_{r+1} = {}^{n+1}C_{r+1}$

Each of these can be proved quite simply by considering the form of ${}^nC_r$. For example, properties 1 and 2 are immediately evident from the definition of ${}^nC_r$. Property 3 can be proved as follows:

$$\begin{aligned}
{}^{n+1}C_1 - {}^nC_1 &= (n+1)!/\{[(n+1)-1]!1!\} - n!/\{[n-1]!1!\} \\
&= (n+1)!/n! - n!/[n-1]! \\
&= (n+1) - n \text{ since } (n+1)! = (n+1)n! \text{ and } n! = n[(n-1)!] \\
&= 1
\end{aligned}$$

The proof of property 4 is a little more involved and is considered as an exercise.

Other properties are not so simply proven. For example

5. $\sum_{r=0}^{n} {}^{n}C_{r} = 2^{n}$

6. $\sum_{r=0}^{n} (-1)^{r}.{}^{n}C_{r} = 0$

7. $\sum_{r=0}^{n} 2^{r}.{}^{n}C_{r} = 3^{n}$

To prove these and similar properties we need a little more information.

Worked example 4.23

Show that ${}^{n}C_{0} = 1$.

Solution: ${}^{n}C_{0} = n!/[(n - 0)!0!] = n!/[n! \times 1] = 1$ because we have defined $0! = 1$.

Worked example 4.24

Show that
(a) ${}^{5}C_{3} + {}^{5}C_{4} = {}^{6}C_{4}$
(b) ${}^{8}C_{5} + {}^{8}C_{5+1} = {}^{8+1}C_{5+1}$
(c) ${}^{n}C_{r} + {}^{n}C_{r+1} = {}^{n+1}C_{r+1}$

Solution:
(a) $\begin{aligned}[t] {}^{5}C_{3} + {}^{5}C_{4} &= 5!/(3!2!) + 5!/(1!4!) \\ &= (5!/[4!2!])(4 + 2) \\ &= (5! \times 6)/(4!2!) \\ &= 6!/(4!2!) \\ &= {}^{6}C_{4} \end{aligned}$

(b) $\begin{aligned}[t] {}^{8}C_{5} + {}^{8}C_{5+1} &= 8!/(3!5!) + 8!/[(8 - [5 + 1])!(5 + 1)!] \\ &= (8!/[3!(5 + 1)!])((5 + 1) + 3) \\ &= (8! \times 9/[3!(5 + 1)!] \\ &= 9!/[3!(5 + 1)!] \\ &= (8 + 1)!/([(8 + 1) - (5 + 1)]!(5 + 1)!) \\ &= {}^{8+1}C_{5+1} \end{aligned}$

(c) $\begin{aligned}[t] {}^{n}C_{r} + {}^{n}C_{r+1} &= n!/[(n - r)!r!] + n!/[(n - [r + 1])!([r + 1]!)] \\ &= (n!/[(n - r)!(r + 1)!])([r + 1] + [n - r]) \\ &= (n!/[(n - r)!(r + 1)!])(n + 1) \\ &= (n + 1)!/[(n - r)!(r + 1)!] \\ &= (n + 1)!/([[n + 1] - [r + 1])!(r + 1)!] \\ &= {}^{n+1}C_{r+1} \end{aligned}$

Exercises

4.23 Show that
(a) $^5C_3 = {}^5C_2$
(b) $^8C_5 = {}^8C_{8-5}$
(c) $^nC_r = {}^nC_{n-r}$

4.24 Show that
(a) $^5C_1 - {}^4C_1 = 1$
(b) $^{8+1}C_1 - {}^8C_1 = 1$
(c) $^{n+1}C_1 - {}^nC_1 = 1$

Pascal's triangle

The following triangle of numbers is known as Pascal's triangle:

Column						
0	1	2	3	4	5	6

Row

0	1						
1	1	1					
2	1	2	1				
3	1	3	3	1			
4	1	4	6	4	1		
5	1	5	10	10	5	1	
6	1	6	15	20	15	6	1

where the dots indicate that the triangle can be built up indefinitely.

The triangle consists of a number of rows and columns, and at the intersection of each is located a number. The value of each number is equal to the number of routes from the apex of the triangle to the location of the number if travel is restricted to vertically down or obliquely to the right as indicated by the lines. Because it is only possible to arrive at a given location from either the location directly above or immediately above and to the left, the number at a given location is equal to the sum of the two numbers directly above and immediately above and to the left. Denoting the number at the intersection of the nth row with the rth column by

$$[n, r]$$

where $n \geq 0$ and $r \leq n$ then

$$[n, r] + [n, r + 1] = [n + 1, r + 1]$$

For example

$$
\begin{aligned}
[2, 0] + [2, 1] &= [2, 0] + [2, 0 + 1] \\
&= 1 + 2 \\
&= 3 \\
&= [3, 1] \\
&= [2 + 1, 0 + 1]
\end{aligned}
$$

Notice the similarity of form between the property

$$[n, r] + [n, r + 1] = [n + 1, r + 1]$$

and

$$^{n}C_{r} + {}^{n}C_{r+1} = {}^{n+1}C_{r+1}$$

It is no accident that the forms of these two equations are similar because, as we shall see, they both express the same facts.

Binomial expansions

A pair of numbers added together and then raised to a power is called a binomial. For example

$$(a + b)^2$$

the square of $a + b$ is a second-order binomial. By evaluating the product we obtain what is called the expansion of the binomial:

$$
\begin{aligned}
(a + b)^2 &= (a + b)(a + b) \\
&= (a + b)a + (a + b)b \\
&= aa + ba + ab + bb \\
&= a^2 + 2ab + b^2
\end{aligned}
$$

and

$$
\begin{aligned}
(a + b)^3 &= (a + b)^2(a + b) \\
&= (a^2 + 2ab + b^2)(a + b) \\
&= (a^2 + 2ab + b^2)a + (a^2 + 2ab + b^2)b \\
&= a^2a + 2aba + b^2a + a^2b + 2abb + b^2b \\
&= a^3 + 3a^2b + 3ab^2 + b^3
\end{aligned}
$$

Notice that the coefficients of the various terms in these expansions are identical to the numbers in Pascal's triangle. Indeed, the coefficients in the expansion of the second-order binomial are located on row 2:

$$
(a + b)^2 = \sum_{r=0}^{2} [2, r]a^{2-r}b^r
$$

and of the third-order binomial on row 3:

$$
(a + b)^3 = \sum_{r=0}^{3} [3, r]a^{3-r}b^r
$$

Consequently, we now have a method of describing the expansion of the general nth-order binomial in terms of the numbers in Pascal's triangle, namely

$$
(a + b)^n = \sum_{r=0}^{n} [n, r]a^{n-r}b^r
$$

Pascal's triangle and the combinatorial coefficients

We have already seen that the numbers in Pascal's triangle possess the same property of form as the combinatorial coefficients, namely

$$
[n, r] + [n, r + 1] = [n + 1, r + 1]
$$

and

$$
{}^nC_r + {}^nC_{r+1} = {}^{n+1}C_{r+1}
$$

We now wish to prove that the numbers in Pascal's triangle are indeed appropriate combinatorial coefficients. We prove this fact by induction.

Proof by induction

Assume that all the numbers on the nth row of Pascal's triangle can be identified with the appropriate combinatorial coefficients. That is

$$
[n, r] = {}^nC_r \text{ for } 0 \leq r \leq n
$$

Because

$$[n + 1, 0] = 1 = {}^{n+1}C_0$$

and

$$[n + 1, n + 1] = 1 = {}^{n+1}C_{n+1}$$

and because

$$[n, r] + [n, r + 1] = [n + 1, r + 1]$$

and

$${}^{n}C_r + {}^{n}C_{r+1} = {}^{n+1}C_{r+1}$$

then all the numbers on the $(n + 1)$th row of Pascal's triangle can also be identified with the appropriate combinatorial coefficients. Finally

$$[0, 0] = 1 = {}^{0}C_0$$

therefore the assumption is correct for all n.

Consequently we can write the expansion of the nth-order binomial as

$$(a + b)^n = \sum_{r=0}^{n} {}^{n}C_r a^{n-r} b^r$$

This now permits us to prove the validity of the last three properties of the combinatorial coefficients mentioned.

If $a = 1$ and $b = 1$ then

$$(1 + 1)^n = \sum_{r=0}^{n} {}^{n}C_r 1^{n-r} 1^r$$

that is

$$2^n = \sum_{r=0}^{n} {}^{n}C_r$$

In other words, the cardinality of the power set of a set with cardinality n is 2^n.

If $a = 1$ and $b = -1$ then

$$(1 - 1)^n = \sum_{r=0}^{n} {}^{n}C_r 1^{n-r} (-1)^r$$

that is

$$0 = \sum_{r=0}^{n} (-1)^r \cdot {}^nC_r$$

which is property 6 proved and finally, if $a = 1$ and $b = 2$, then

$$(1 + 2)^n = \sum_{r=0}^{n} {}^nC_r 1^{n-r} 2^r$$

that is

$$3^n = \sum_{r=0}^{n} 2^r \cdot {}^nC_r$$

which proves property 7.

Worked example 4.25

Write down the coefficients in row 7 of Pascal's triangle and show that they add up to 2^7.

Solution: The coefficients in row 6 of Pascal's triangle are:

1, 6, 15, 20, 15, 6, 1

so the coefficients in row 7 are

1, (1 + 6) = 7, (6 + 15) = 21, (15 + 20) = 35, (20 + 15) = 35, (15 + 6) = 21, (6 + 1) = 7, 1

These coefficients add up to $128 = 2^7$.

Worked example 4.26

Expand $(a + b)^4$ and show that the coefficients match up to those in Pascal's triangle.

Solution:
$$\begin{aligned}
(a + b)^4 &= (a + b)^3(a + b) \\
&= (a^3 + 3a^2b + 3ab^2 + b^3)(a + b) \\
&= a^3a + 3a^2ba + 3ab^2a + b^3a + a^3b + 3a^2bb + 3ab^2b + b^3b \\
&= a^4 + 4a^3b + 6a^2b^2 + 4ab^3 + b^4
\end{aligned}$$

Worked example 4.27

Show that

(a) $\sum_{r=0}^{3} {}^3C_r = 2^3$

(b) $\sum_{r=0}^{n} a^r.{}^nC_r = (a + 1)^n$

Solution:

(a) $\sum_{r=0}^{3} {}^3C_r$ $= {}^3C_0 + {}^3C_1 + {}^3C_2 + {}^3C_3$
$= 3!/(3!0!) + 3!/(2!1!) + 3!/(1!2!) + 3!/(0!3!)$
$= 1 + 3 + 3 + 1$
$= 8$
$= 2^3$

(b) $\sum_{r=0}^{n} a^r.{}^nC_r = \sum_{r=0}^{n} (1)^{n-r} a^r.{}^nC_r = (1 + a)^n$

Exercises

4.25 Write down the coefficients in row 8 of Pascal's triangle and show that they add up to 2^8.

4.26 Expand $(a + b)^5$ and show that the coefficients match up to those in Pascal's triangle.

4.27 Show that

(a) $\sum_{r=0}^{4} 3^r.{}^4C_r = 4^4$

(b) $\sum_{r=0}^{n} (1/n)^r.{}^nC_r = (1 + 1/n)^n$

and that as n increases this value approaches the irrational number e which, to five decimal places, is 2.71828.

Part Three

Further Logic

By now you will realize the power of the logic constructions that we have discussed so far. We have found that many of our uses of the ordinary English language can be analysed, deconstructed and then reconstructed to enable them to be generalized within a well-defined mathematical framework that is accessible to everyone. There are still, however, many aspects of English that we have yet to consider: how do we cater for words such as **some**, **all**, **each** and **every**? To address these issues we must recognize that, while we have laid the foundations of logic in the propositional calculus, we now need to build upon them. We need to extend our ability to discuss the truth or falsity of our assertions to situations in which an assertion can only be assigned a truth value after certain external criteria have been applied. Furthermore, having made such assertions we need to know that what we say is correct – we need to be able to prove our assertions to be correct and, even more importantly, we need to be able to demonstrate that our conclusions are consistent with the mathematical structures we are creating.

Chapter 5

Predicate calculus

OBJECTIVES

When you have completed this chapter you will be able to:

☐ distinguish between a variable and a predicate in an open sentence;

☐ demonstrate the uses of the universal and existential quantifiers;

☐ negate propositions containing the universal and existential quantifiers.

You are faced with the problem of putting a 3000-piece jigsaw together. Where on earth do you start? You look at the picture on the box lid. Here is a large patch of sky, here is a lake, here in the top right-hand corner is a tree full of autumn tints and here in the centre is a herd of cattle. Jigsaws go together by joining one piece to another but you know from experience that such a large jigsaw requires a systematic method to get you off the ground. Look for the border pieces; if you can put all the border pieces together then you can start working inwards. Every border piece has a straight edge. Some have two – these are the corner pieces. Now the jigsaw is starting to take shape.

Completing a jigsaw requires an ability to argue from the general to the particular. All sky pieces have light blue on them. This particular piece has light blue on it therefore it is a sky piece. All border pieces have a straight edge. This piece has a straight edge therefore it is a border piece. In each case we test a proposition against a condition or criterion to find its truth value.

So far we have considered propositions to which a truth value can be immediately assigned. Such propositions are very specific and, as a consequence, have a limited application. In reality, we need to be able to handle propositions whose truth value can only be ascertained when they have been tested against some external criterion. This piece of jigsaw is a border piece. How do I know whether this statement is true or false? I test it against the criterion that border pieces have a straight edge. If the piece has a straight edge then the statement is true. If not then the statement is false. To cater for this situation we need to extend the propositional calculus to enable us to discuss propositions that are sometimes true and sometimes false.

Propositions and their predicates

Predicates and open sentences
Every proposition is a sentence containing a **subject** and a phrase that
describes the subject. For example, the proposition

   Bill is a student

is a sentence with subject 'Bill' – the subject being what the sentence is
about. The remainder of the sentence is the phrase

   is a student

which describes Bill. This descriptive phrase is called a **predicate**.
   A predicate is the descriptive part of a proposition, and the truth value of
a proposition is decided by applying that description to the subject of the
proposition. Consequently, while it might seem at first sight that it is the
description that determines the truth value of a proposition, it is not. It is
the subject that determines the truth value because the subject either fits the
description or it does not. The truth or falsity of the proposition

   Bill is a student

depends upon Bill: he either is or is not a student.
   This separation of a proposition into subject and predicate is a most useful device that permits the creation of a more extensive logical object than a
mere simple proposition. By replacing the subject of a proposition with a
variable we can construct an **open sentence**. For example

   x is a student

is an open sentence – open in the sense that until we replace x with a
specific value we are unable to determine the truth value of the sentence.
   The flexibility permitted by the introduction of variables enables more
complicated statements to be introduced into logic. For example

   $$x^2 - 5x + 6 = 0 \text{ and } x^2 + y^2 + z^2 = 25$$

are both open sentences that are capable of being incorporated in a proposition.

*Conclusion: Every open sentence contains at least one variable and its
associated predicate.*

Worked example 5.1

Identify the subject and predicate in each of the following propositions:
(a) The dog bit the man.
(b) Most people like sugar.
(c) Happiness is dog shaped.
(d) Today the weather is wet and windy.

Solution:
(a) Subject: The dog
     Predicate: bit the man
(b) Subject: Most people
     Predicate: like sugar
(c) Subject: Happiness
     Predicate: is dog shaped
(d) Subject: Today
     Predicate: the weather is wet and windy

Worked example 5.2

Which of the following is a proposition and which is an open sentence?
(a) Joan is married to John.
(b) She is married to John.
(c) Five is greater than four.
(d) The number is greater than four.

Solution:
(a) Proposition.
(b) Not a proposition because until 'She' is identified it is not possible to assign a truth value.
(c) Proposition.
(d) Not a proposition because until 'the number' is identified it is not possible to assign a truth value.

Worked example 5.3

Convert each of the following propositions into an open sentence by appropriate choice of variable:
(a) The cat is black.
(b) Jack is at least 6 feet tall.
(c) $8 > 4$
(d) The modulus of an integer is a positive integer.

Solution:
(a) x is black **or** The cat is x.
(b) x is at least 6 feet tall **or** Jack is at least x tall.
(c) $x > 4$ **or** $8 > x$
(d) $|n| = n$ if $n \geq 0$
       $= -n$ if $n < 0$

Exercises

5.1 Identify the subject and predicate in each of the following propositions:
(a) John is a computer student.
(b) Some of the people can be fooled some of the time.
(c) Procrastination is the thief of time.
(d) This is a statement.

5.2 Which of the following is a proposition and which is an open sentence?
(a) Jack Spratt could eat no fat.
(b) He could eat no fat.
(c) x is a letter of the alphabet.
(d) x is a number greater than 7.

5.3 Convert each of the following propositions into an open sentence by appropriate choice of variable:
(a) The front door is painted green.
(b) Anne could do better.
(c) The bus will arrive at noon.
(d) The square of an integer is larger than the integer.

Truth values

Every open sentence contains at least one variable, and the process of determining the truth value of the sentence begins by substituting specific values for the variable or variables. This, of course, begs the question: What are the valid variable values? To answer this question we must indicate the set of variable values which can be substituted into the sentence to enable its truth value to be determined in each case. For example, the open sentence

$$x^2 = 1$$

carries with it the tacit assumption that the x-values are numbers. However, tacit assumptions are insufficient, we must be explicit. We shall rewrite this open sentence as

Remember that we use $\mathcal{R}$ to denote the set of real numbers.

$$x^2 = 1 \text{ where } x \in \mathcal{R}$$

to indicate that x is a real number. Next we substitute specific real numbers for the variable and deduce the truth values of the resulting propositions.

It is a simple matter to show that substitution of the x-values

$$x = 1 \text{ and } x = -1$$

convert this open sentence into a true proposition:

$(1)^2 = 1$ where $1 \in \mathcal{R}$ is true
$(-1)^2 = 1$ where $-1 \in \mathcal{R}$ is true

Substituting any other real number will convert the open sentence into a false proposition. Here we see the essence of what is called the **predicate calculus** – the substitution of variable values into open sentences and the determining of the truth value of the resulting proposition.

Notation
An open sentence will be indicated by a capital letter and the associated variable or variables by lower-case letters. For example

$P(x)$

is an open sentence involving a single variable x and

$Q(x, y, z)$

is an open sentence containing the three variables x, y and z. In either case the open sentence only becomes a proposition when values are substituted for the variables. For example, if $P(x, y)$ represents the open sentence

$$x^2 + y^2 = 4, \ x, y \in \mathcal{R}$$

then the proposition

$P(2, 0)$

is true, whereas the proposition

$P(1, 1)$

is false.

Conclusion: Associated with every variable in an open sentence is a set of variable values. Substituting values from the sets for the appropriate variables converts the open sentence into a proposition.

Worked example 5.4

Each of the following open sentences is accompanied by a set of variable values. Which values convert the open sentence into a true proposition?
(a) $|x| = x$, $\{-2, -1, 0, 1, 2\}$
(b) x is a vowel, $\{a, b, c, d, e, f\}$
(c) $x^2 - y^2 \leq 1$, $\{(0, 0), (-1, -1), (-2, 1), (3, -1), (4, 1)\}$
(d) $x + 2y - 3z = 4$, $\{(0, 1, 2), (0, 2, 0), (4, 3, 2), (-5, 3, -1), (1, 1, 1)\}$

Solution:
(a) $\{0, 1, 2\}$
(b) $\{a, e\}$
(c) $\{(0, 0), (-1, -1)\}$
(d) $\{(0, 2, 0), (4, 3, 2), (-5, 3, -1)\}$

Exercise

5.4 Each of the following open sentences is accompanied by a set of variable values. Which values convert the open sentence into a true proposition?
(a) $x^2 \leq x$, $\{-0.2, -0.1, 0.0, 0.1, 0.2\}$
(b) x is a fruit, $\{$apple, tomato, potato, pear, mango, runner bean$\}$
(c) $x^2 + y^2 \geq 0.1$, $\{(0, 0), (-0.1, -0.3), (-0.2, 0.7), (0.3, -0.1), (0.4, -0.45)\}$
(d) $5p - 10q + 4r \leq -6$, $\{(0, 1, 2), (0, 2, 0), (4, 3, 2), (-5, 3, -1), (1, 1, 1)\}$

Universality
Some open sentences convert to a true proposition **for all possible value substitutions** of the variable or variables. For example, the open sentence

$$x^2 - 1 = (x + 1)(x - 1), \ x \in \mathcal{R}$$

is converted to a true proposition no matter which real number value is substituted for x. We say that this open sentence converts to a **universally** true proposition and write this fact as the **quantified proposition**:

for all $x \in \mathcal{R}, \ x^2 - 1 = (x + 1)(x - 1)$

The phrase **for all** signifies that this proposition is universally true – true for every substituted value of x. We use the symbol $\forall$ to stand for the phrase **for all** and write the proposition as

$$\forall x \in \mathcal{R}, \; x^2 - 1 = (x+1)(x-1)$$

The symbol $\forall$ is called the **universal quantifier** and the fact that an open sentence P(x) converts to a universally true proposition is written as

$$\forall x \in X, \; P(x)$$

There are a number of English alternatives for the phrase **for all**. For example, **every**, **each** or, simply, **all**.

Conclusion: The universal quantifier is used to describe the fact that an open sentence is converted to a true proposition for all substituted values of the variables. The proposition

$$\forall x \in X, \; P(x)$$

is referred to as a universally quantified proposition.

Worked example 5.5

Determine which of the following open sentences are universally quantified propositions and write the proposition using the universal quantifier:
(a) All jockeys are short.
(b) Some jockeys are short.
(c) There are short people who are jockeys.
(d) Each short person is a jockey.

Solution: Only (a) and (d) are universally quantified propositions.
(a) $\forall x, x$ is short where $x \in \{j: j$ is a jockey$\}$
(b) This is not a universally quantifiable proposition.
(c) This is not a universally quantifiable proposition.
(d) $\forall x, x$ is a jockey where $x \in \{s: s$ is a short person$\}$

Worked example 5.6

Write each of the following sentences symbolically:
(a) Every book has a hard cover.
(b) All summer days are sunny.
(c) Each apple in the tray is green.
(d) Negative numbers are less than positive numbers.

Solution:
(a) $\forall x, x$ has a hard cover where $x \in \{b: b$ is a book$\}$
(b) $\forall x, x$ is sunny where $x \in \{s: s$ is a summer day$\}$
(c) $\forall x, x$ is green where $x \in \{a: a$ is an apple in the tray$\}$
(d) $\forall x, x < 0$ where $x \in \{n: n$ is a negative number$\}$

Exercises

5.5 Determine which of the following open sentences convert to a universally quantified proposition and write the proposition using the universal quantifier:
(a) All trees have green leaves.
(b) Some trees have green leaves.
(c) Every tree has green leaves.
(d) There is a time and a place for everything.

5.6 Write each of the following sentences symbolically:
(a) Every dog has his day.
(b) Each flower has at least three petals.
(c) None of the oranges are ripe.
(d) All English words possess at least one vowel.

Existence
Sometimes it may be sufficient to indicate that a given open sentence converts to a true proposition **for at least one substituted value** of the variable or variables. For example, the open sentence

$$x + 1 = 0, \ x \in \mathcal{R}$$

converts to a true proposition only for the single substituted value $x = -1$. In this case we can say that

there exists a value of $x \in \mathcal{R}$ such that $x + 1 = 0$

The phrase **there exists** indicates that there is at least one value that can be substituted for x for which this open sentence converts to a true proposition. It may be the case that more than one value of x can be substituted to convert the open sentence into a true proposition but that is of no account: we are concerned with the fact that there is **at least one**. We use the symbol $\exists$ to stand for the phrase **there exists** and write the quantified proposition as

As with the universal **for all** so with the existential **there exists** there are a number of English alternatives, for example **there is** or **some**.

$$\exists x \in \mathcal{R}, \ x + 1 = 0$$

The symbol $\exists$ is called the **existential quantifier**. The fact that an open sentence $P(x)$ converts to a true proposition for at least one substituted value of the variable x is written as

$$\exists x \in X, P(x)$$

Conclusion: The existential quantifier is used to describe the fact that an

*open sentence is converted to a true proposition for at least one of the sub-
stituted values of the variables. The proposition*

$$\exists x \in X, P(x)$$

is referred to as an existentially quantified proposition.

Worked example 5.7

Determine which of the following open sentences are existentially quantified propositions and write the proposition using the existential quantifier:
(a) Some days are better than others.
(b) Each item was bought in the local store.
(c) There is a bank where the wild thyme grows.
(d) If x is a natural number then $\sqrt{x}$ is not a natural number.

Solution: Only (a) and (c) are existentially quantified propositions.
(a) $\exists x \in \{x: x \text{ is a day}\}, x$ is better than other days
(c) $\exists x \in \{x: x \text{ is a bank}\}, x$ is where the wild thyme grows

Worked example 5.8

Write each of the following sentences symbolically:
(a) Some handwriting specimens are illegible.
(b) One of the trains will travel south.
(c) I have a copy of that file on at least one of my disks.
(d) There are numbers that satisfy the equation $x^2 = -1$.

Solution:
(a) $\exists x \in \{x: x \text{ is a handwriting specimen}\}, x$ is illegible
(b) $\exists x \in \{x: x \text{ is a train}\}, x$ will travel south
(c) $\exists x \in \{x: x \text{ is my disk}\},$ I have a copy of that file on x
(d) $\exists x \in \{x: x \text{ is a number}\}, x^2 = -1$

Exercises

5.7 Determine which of the following open sentences convert to an existentially quantified proposition and write the proposition using the
existential quantifier:
(a) Sometimes I go to Edinburgh by train.
(b) Each database record has a Name field.
(c) There will come a time when all will be clear.

(d)   A number x exists where x^2 is not an irrational number.

5.8   Write each of the following sentences symbolically:
   (a)   Some buses are always late.
   (b)   One day your boat will come in.
   (c)   At least one of the students scored full marks.
   (d)   There is a copy of this book in your library.

Negating quantified propositions

Negating universally quantified propositions
The universally quantified proposition

   All men have their hair cut by a barber

can be negated to give the proposition

   Not all men have their hair cut by a barber

This proposition is equivalent to

   There is a man who does not have his hair cut by a barber

To negate the quantifier all it is not necessary to state exactly how many do not have their hair cut by a barber, it is sufficient to state that there is at least one who does not.
   Symbolically we have to negate

This forms the basis of **disproof by exception**, where a universal statement is proved incorrect by finding a single example for which it does not hold true.

   $\forall x \in \{\text{men}\}$, x has his hair cut by a barber

This we do using the symbol $\neg$ to denote negation

   $\neg(\forall x \in \{\text{men}\}$, x has his hair cut by a barber$)$

and we have just seen that this is equivalent to

   $\exists x \in \{\text{men}\}$, x does not have his hair cut by a barber

In summary, the negation of the universal quantifier applied to $P(x)$ is logically equivalent to the existential quantifier applied to the negation of $P(x)$:

   $\neg[\forall x \in X, P(x)] \equiv \exists x \in X, \neg P(x)$

Conclusion: To negate a proposition containing the universal quantifier and an open sentence change the universal quantifier to an existential quantifier and negate the open sentence.

Worked example 5.9

Negate each of the following propositions:
(a) All apples are green.
(b) Every day is a sunny day.
(c) Everybody in the region speaks French.
(d) Each of these articles represents the result of a fault in the assembly line.

Solution:
(a) Not all apples are green **or** There is an apple that is not green.
(b) Not every day is a sunny day **or** There is a day that is not sunny.
(c) Not everybody in the region speaks French **or** There is a person in the region who does not speak French.
(d) Not all of these articles represents the result of a fault in the assembly line **or** There is an article that does not represent a fault in the assembly line.

Worked example 5.10

Convert the following English statements into universally quantified propositions in symbolic form, negate them and then convert the negations into English statements:
(a) All apples are green.
(b) Every day is a sunny day.
(c) Everybody in the region speaks French.
(d) Each of these articles represents the result of a fault in the assembly line.

Solution:
(a) $\forall x \in \{\text{apples}\}, x$ is green
     $\neg(\forall x \in \{\text{apples}\}, x$ is green$)$
     $\exists x \in \{\text{apples}\}, x$ is not green
     There is an apple that is not green.
(b) $\forall x \in \{\text{days}\}, x$ is sunny
     $\neg(\forall x \in \{\text{days}\}, x$ is sunny$)$
     $\exists x \in \{\text{days}\}, x$ is not sunny
     There is a day that is not sunny.
(c) $\forall x \in \{\text{people in the region}\}, x$ speaks French

¬(∀x ∈ {people in the region}, x speaks French)
∃x ∈ {people in the region}, x does not speak French
There is a person in the region who does not speak French.

(d)   ∀x ∈ {articles}, x represents a fault in the assembly line
¬(∀x ∈ {articles}, x represents a fault in the assembly line)
∃x ∈ {articles}, x does not represent a fault in the assembly line
There is an article that does not represent a fault in the assembly line.

Exercises

5.9   Negate each of the following propositions:
   (a)   All the computers are PCs.
   (b)   Everybody likes surprises.
   (c)   Every house is built in brick.
   (d)   Each one of the voters disagreed with the candidate.

5.10 Convert the following English statements into universally quantified propositions in symbolic form, negate them and then convert the negations into English statements:
   (a)   All the computers are PCs.
   (b)   Everybody likes surprises.
   (c)   Every house is built in brick.
   (d)   Each one of the voters disagreed with the candidate.

Negating existentially quantified propositions
The existentially quantified proposition

There is a planet that is made of green cheese

can be negated to give the proposition

There is not a planet made of green cheese

This proposition is equivalent to

No planets are made of green cheese

or

Planets are not made of green cheese

To negate the quantifier **there is a** it is necessary to state that **all** planets are not made of green cheese.
   Symbolically we have to negate

$\exists x \in \{\text{planets}\}, x$ is made of green cheese

This we do using the symbol $\neg$ to denote negation:

$\neg(\exists x \in \{\text{planets}\}, x$ is made of green cheese)

and we have just seen that this is equivalent to

$\forall x \in \{\text{planets}\}, x$ is not made of green cheese

In summary, the negation of the existential quantifier applied to P(x) is logically equivalent to the universal quantifier applied to the negation of P(x):

$$\neg[\exists x \in X, P(x)] \equiv \forall x \in X, \neg P(x)$$

Conclusion: To negate a proposition containing the existential quantifier and an open sentence change the existential quantifier to a universal quantifier and negate the open sentence.

Worked example 5.11

Negate each of the following propositions:
(a) There is a mountain in Wales as high as 4000 feet.
(b) Some of these are better than those.
(c) There is a house in New Orleans they call the Rising Sun.
(d) Somewhere over the rainbow skies are blue.

Solution:
(a) There is not a mountain in Wales as high as 4000 feet **or** No mountains in Wales are as high as 4000 feet.
(b) None of these are better than those.
(c) There is no house in New Orleans they call the Rising Sun.
(d) Nowhere over the rainbow are skies blue.

Worked example 5.12

Convert the following English statements into existentially quantified propositions in symbolic form, negate them and then convert the negations into English statements:
(a) There is a mountain in Wales as high as 4000 feet.
(b) Some of these are better than those.
(c) There is a house in New Orleans they call the Rising Sun.
(d) Somewhere over the rainbow skies are blue.

Solution:

(a) Writing this statement symbolically we have

[∃x ∈ {mountains in Wales}, x is as high as 4000 feet]

Negating gives

¬[∃x ∈ {mountains in Wales}, x is as high as 4000 feet]
≡ ∀x ∈ {mountains in Wales}, ¬(x is as high as 4000 feet)

This translates into English as: There are no mountains in Wales as high as 4000 feet.

(b) Writing this statement symbolically we have

[∃x ∈ {items}, x is better than those]

Negating gives

¬[∃x ∈ {items}, x is better than those]
≡ ∀x ∈ {items}, ¬(x is better than those)

This translates into English as: None of these are better than those.

(c) Writing this statement symbolically we have

[∃x ∈ {houses in New Orleans}, x is called the Rising Sun]

Negating gives

¬[∃x ∈ {houses in New Orleans}, x is called the Rising Sun]
≡ ∀x ∈ {houses in New Orleans}, ¬(x is called the Rising Sun)

This translates into English as: No house in New Orleans is called the Rising Sun.

(d) Writing this statement symbolically we have

[∃x ∈ {places over the rainbow}, in x the skies are blue]

Negating gives

¬[∃x ∈ {places over the rainbow}, in x the skies are blue]
≡ ∀x ∈ {places over the rainbow}, ¬(in x the skies are blue)

This translates into English as: Nowhere over the rainbow are skies blue.

Exercises

5.11 Negate each of the following propositions:
  (a) There is a gold mine in those hills.
  (b) Someday my luck will change.
  (c) There is a hole in the road.
  (d) Sometime I shall go there.

5.12 Convert the following English statements into existentially quantified propositions in symbolic form, negate them and then convert the negations into English statements:
(a) There is a gold mine in those hills.
(b) Someday my luck will change.
(c) There is a hole in the road.
(d) Sometime I shall go there.

Multiply quantified propositions

The universal and existential quantifiers can be coupled together to form a multiply quantified proposition. For example, if I were to say that

For every day of the week there is a time when I eat lunch

This can be written symbolically as

$\forall x \in \{$days of week$\}$, $\exists y \in \{$times$\}$, on x I eat lunch at y

As another example, if I were to say that

There is a time when I each lunch every day of the week

This can be written symbolically as

$\exists y \in \{$times$\}$, $\forall x \in \{$days of week$\}$, on x I eat lunch at y

It is important to realize that these two multiply quantified propositions have different meanings. The first states that I have lunch every day, whereas the second one states that I have lunch every day at the same time.

Conclusion: A multiply quantified proposition is a proposition containing more than one quantifier.

Worked example 5.13

Write each of the following statements symbolically as a multiply quantified proposition:
(a) Every dog has his day.
(b) Everyone has a day of the week that is special.
(c) There is a time when everyone should come to the aid of the party.
(d) You can fool all of the people some of the time.

Solution:
(a) $\forall x \in \{$dogs$\}$, $\exists y \in \{$days$\}$, x has his y
(b) $\forall x \in \{$people$\}$, $\exists y \in \{$days of the week$\}$, y is a special day to x
(c) $\exists x \in \{$times$\}$, $\forall y \in \{$people$\}$, y should come to the aid of the party at x
(d) $\forall x \in \{$people$\}$, $\exists y \in \{$times$\}$, x can be fooled at time y

Worked example 5.14

Translate each of the following multiply quantified propositions into English statements:

(a) $\forall x \in \mathcal{R}, \exists y \in \mathcal{R}, y < x$

(b) $\forall x \in \{$European states$\}$, $\exists y \in \{$people$\}$, y will be President of the Union when it is x's turn

(c) $\forall x \in \{$items$\}$, $\exists y \in \{$places$\}$, y is x's place

(d) $\exists x \in \{$people$\}$, $\forall y \in \{$times$\}$, x can be fooled at time y

Solution:

(a) There is no smallest real number.

(b) In every member state of the European Union there is somebody who will be President of the Union when their country's turn comes around.

(c) To every thing there is a place.

(d) You can fool some of the people all of the time.

Exercises

5.13 Write each of the following statements symbolically as a multiply quantified proposition:

    (a) Every English word contains at least one vowel.

    (b) No-one can be right all of the time.

    (c) Someday everybody will arrive together.

    (d) There is at least one student who takes all five courses.

5.14 Translate each of the following multiply quantified propositions into English statements:

    (a) $\forall x \in \{$Waverley novels$\}$, $\exists y \in \{$people$\}$, y has not read x

    (b) $\forall x \in \{$facts$\}$, $\exists y \in \{$people$\}$, y knows fact x

    (c) $\forall x \in \{$villages$\}$, $\exists y \in \{$maps$\}$, x is found on y

    (d) $\exists x \in \{$library books$\}$, $\forall y \in \{$pages$\}$, x has a diagram on y

---

Semantics versus syntax

A study of **semantics** deals with the meaning of language relating to the meaning and implications of words, whereas a study of **syntax** deals with the grammatical arrangement of words and sentences showing their connection and structure.

If we consider the compound proposition

Today is Monday and it is sunny

then we understand this to have the same meaning as the compound proposition

It is sunny and today is Monday

By appealing to meaning in this way we are considering the **semantics** of the compound proposition. If, on the other hand, we state that for any two propositions p and q

$$p \wedge q \equiv q \wedge p$$

we are stating that the structure of the compound proposition on the left is logically equivalent to the structure of the compound proposition on the right. By appealing to the structure or **form** of the compound proposition in this way we are dealing with the syntax of the logic.

Negating multiply quantified propositions

Negating a multiply quantified proposition that is given in the form of an English sentence can sometimes be a daunting process if it is attempted from the meaning, or semantics, of the proposition. Because of the inherent ambiguity of the English language it can be a difficult process to properly divine the meaning of a multiply quantified proposition before the negation is even considered. To overcome problems of this nature it is advisable to convert the English statement into symbolic form and then to negate that form.

To negate a multiply quantified proposition in symbolic form it is necessary to bear in mind the method employed to negate a proposition containing a single quantifier. The rule is

Switch the quantifier and negate the open sentence

For example, how do we negate the following compound proposition?

Every course has a component that is an elective

One problem with trying to negate this proposition is to decide which possibility is actually the negation. Is it

Every course has a component that is not an elective

or is it

Not every course has a component that is an elective

or is it

Not every course has a component that is not an elective

and if it is the last what exactly does it mean?

To avoid this sort of problem write the original proposition as

For all courses x there exists a component y such that the component y of course x is an elective

This can be written symbolically as

$$\forall x \in X, \exists y \in Y, P(x, y)$$

where $X = \{\text{courses}\}$, $Y = \{\text{components}\}$ and $P(x, y) = $ component y of course x is an elective.

We first write the negation as

$$\neg[\forall x \in X, \exists y \in Y, P(x, y)] = \neg[\forall x \in X, \{\exists y \in Y, P(x, y)\}]$$

Here the bracketed expression

$$\{\exists y \in Y, P(x, y)\}$$

is an open sentence because within the brackets the variable x is not bound to a quantifier. The negation above then becomes, by application of the rule,

$$\neg[\forall x \in X, \{\exists y \in Y, P(x, y)\}] = [\exists x \in X, \neg\{\exists y \in Y, P(x, y)\}]$$

Next, we negate the open sentence, again employing the rule

$$\neg\{\exists y \in Y, P(x, y)\} = \forall y \in Y, \neg P(x, y)$$

Consequently, we see that

$$\neg[\forall x \in X, \{\exists y \in Y, P(x, y)\}] = \exists x \in X, \forall y \in Y, \neg P(x, y)$$

Translating this symbolic proposition we see that

There exists a course x such that for all components y the component y of course x is not an elective

or

Not all courses have an elective component

Following such a systematic procedure makes the negating of a compound proposition a relatively straightforward matter.

Conclusion: Negating a compound proposition is best performed when the proposition is written symbolically. The negation is then achieved by a straightforward application of an extended form of the rule for negating a proposition containing just one quantifier.

Worked example 5.15

Negate the following multiply quantified propositions:
(a) $\forall x \in$ {Waverley novels}, $\exists y \in$ {people}, y has not read x
(b) $\forall x \in$ {facts}, $\exists y \in$ {people}, y knows fact x
(c) $\forall x \in$ {villages}, $\exists y \in$ {maps}, x is found on y
(d) $\exists x \in$ {library books}, $\forall y \in$ {pages}, x has a diagram on y

Solution:
(a) The negation is

   $\exists x \in$ {Waverley novels}, $\forall y \in$ {people}, y has read x

(b) The negation is

   $\exists x \in$ {facts}, $\forall y \in$ {people}, y does not know fact x

(c) The negation is

   $\exists x \in$ {villages}, $\forall y \in$ {maps}, x is not found on y

(d) The negation is

   $\forall x \in$ {library books}, $\exists y \in$ {people}, x has no diagram on y

---

Worked example 5.16

Negate each of the following English statements giving each answer in the form of another English statement:
(a) Every English word contains at least one vowel.
(b) No-one can be right all of the time.
(c) Someday everybody will arrive together.
(d) There is at least one student who takes all five courses.

Solution:
(a) Symbolically this is

   $\forall x \in$ {English words}, $\exists y \in$ {vowels}, x contains y

   The negation is

   $\exists x \in$ {English words}, $\forall y \in$ {vowels}, x does not contain y

   Translated into English this reads: There is an English word that does not contain a vowel.

(b) Symbolically this is

$\forall x \in \{\text{people}\}, \exists y \in \{\text{time}\}, x$ is not right at y

The negation is

$\exists x \in \{\text{people}\}, \forall y \in \{\text{time}\}, x$ is right at y

Translated into English this reads: There is a person who is always right.

(c) Symbolically this is

$\exists x \in \{\text{days}\}, \forall y \in \{\text{people}\}, y$ will arrive on x

The negation is

$\forall x \in \{\text{days}\}, \exists y \in \{\text{people}\}, y$ will not arrive on x

Translated into English this reads: Every day there is a person who will not arrive.

(d) Symbolically this is

$\exists x \in \{\text{students}\}, \forall y \in \{\text{courses}\}, x$ will take y

The negation is

$\forall x \in \{\text{students}\}, \exists y \in \{\text{courses}\}, x$ will not take y

Translated into English this reads: There is a course that will not be taken by all the students.

Exercises

5.15 Negate the following multiply quantified propositions:
   (a) $\forall x \in \mathcal{R}, \exists y \in \mathcal{R}, y < x$
   (b) $\forall x \in \{\text{European states}\}, \exists y \in \{\text{people}\}, y$ will be President of the Union when it is x's turn
   (c) $\forall x \in \{\text{items}\}, \exists y \in \{\text{places}\}, y$ is x's place
   (d) $\exists x \in \{\text{people}\}, \forall y \in \{\text{times}\}, x$ can be fooled at time y

5.16 Negate each of the following English statements giving each answer in the form of another English statement:
   (a) Every dog has his day.
   (b) Everyone has a day of the week that is special.
   (c) There is a time when everyone should come to the aid of the party.
   (d) You can fool all of the people some of the time.

Chapter 6

Proof

OBJECTIVES

When you have completed this chapter you will be able to:

☐ explain the meaning of the word proof and appreciate the need for proofs;

☐ prove the validity of argument forms;

☐ prove arguments and theorems using the Gentzen rules of natural deduction.

In a British court of law the defendant is presumed to be innocent until proven guilty. Guilt is proved by the counsel for the prosecution presenting evidence to the court that demonstrates beyond reasonable doubt that the defendant did in fact commit the actions of which he or she is accused. The process of putting a defendant on trial in such a court is traditionally one of contest. The prosecuting counsel competes with the defending counsel in presenting and debating their cases. The one with the best case wins and, if it is the prosecuting counsel that wins, the defendant is found guilty of the charges so laid before the court – the case for the prosecution is said to have been proved.

Proof in mathematics is another matter entirely. There are similarities in that evidence is accrued to support the argument, but there the similarities end. The process of proving assertions to be true or false is not one of contest between opposing sides, it is one of demonstrating that a stated assertion is consistent with everything that has been said before and that the assertion is a necessary consequence of the mathematical framework within which the proof is derived. There is no such thing as proof beyond reasonable doubt in mathematics, there is only proof without any doubt whatsoever.

Proving assertions

The need for proof
When the Greek philosopher Aristotle attempted to describe the nature of motion he said that all matter was composed of a mixture of four basic elements: earth, air, fire and water. Matter made from a preponderance of earth exhibited a tendency to move downwards and matter made from a preponderance of fire tended to rise. His theories were derived from observation coupled with an intuitive feel about the constitution of matter, and, because no-one had a better theory, his ideas held sway for many centuries. The Egyptian astronomer Ptolemy had a theory that the earth was located at the centre of the universe and that all the observed heavenly bodies rotated about this centre. Both Aristotle's theory of motion and Ptolemy's theory of celestial construction had immediately recognized deficiencies in that they were unable to give an adequate account of the motion of the planets which, when viewed from the earth, have highly convoluted trajectories. Aristotle tried to explain these trajectories by saying that the planets moved the way they did because they were being pushed by angels. Ptolemy's attempt to explain these motions involved the construction of his theory of epicycles, according to which the planets were fixed to the circumferences of circles that were rotating within other circles.

It was not until the fifteenth century that Galileo looked through his new telescope and discovered that indeed the sun was at the centre of the solar system and that the earth and all the planets revolved around the sun in near-circular orbits. His discoveries caused a revolution in scientific thought, not least because when challenged he could point to his telescope and say 'I can prove it. Just look through this telescope'.

This notion of proof is essential to all scientific endeavour. There is no point in making an assertion unless you can prove that what you are saying is true. Furthermore, the ability to prove your assertions to be true must be independent of the person proving them; there must be an accepted consensus of logic that can be employed by anyone in the attempt to prove statements to be true.

In the fifth century BC Pythagoras discovered his now eponymous relationship between the hypotenuse of a right-angled triangle and the other two sides:

> The square on the hypotenuse of a right-angled triangle is equal to the sum of the squares on the other two sides

How he came to discover it is unknown, but what is clear is that it is an established fact and can be proved using accepted ideas of plane geometry. And herein lies the essence of proof. Any statement made is made within a context that is governed by a set of accepted ideas. Proof that the statement is true depends upon an argument that accepts earlier ideas as being true – we call them premises – and then builds upon those ideas using a logical argument to produce a conclusion that the statement is true.

Consider the proof of Pythagoras' theorem beginning with Figure 6.1 of a square of side length a.

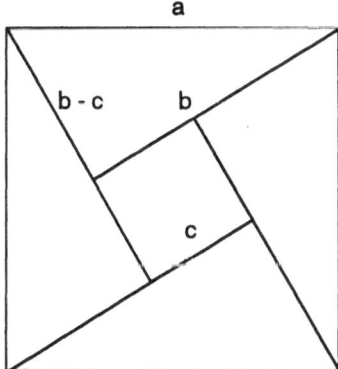

Figure 6.1

The square is constructed from four identical right-angled triangles with a small square of side length c in the middle. The side lengths of each of the triangles are a, b and $b - c$.

Pythagoras' theorem states that

$$a^2 = b^2 + (b - c)^2$$

The proof follows.

Proof

1. The length of the hypotenuse of each triangle is a, which makes the area of the square

    a^2

2. The lengths of the other two sides of each of the triangles are b and $b - c$, giving the area of each triangle as

    $(1/2)b(b - c)$

3. The area of the square is equal to the areas of the four right-angled triangles plus the area of the little square of side c. That is

    $$\begin{aligned}
    a^2 &= 4(1/2)b(b - c) + c^2 \\
    &= 2b^2 - 2bc + c^2 \\
    &= b^2 + (b^2 - 2bc + c^2) \\
    &= b^2 + (b - c)^2
    \end{aligned}$$

    which proves the theorem.

Notice that the statements about area are commonly accepted as true and the use of the algebraic manipulations are commonly accepted as valid. This is very important, because this forms the context within which the theorem is stated and proved; all proofs are constructed using commonly accepted truths. Indeed, the word theorem itself refers to any statement that can be **derived** (by which is meant shown to be true) from an accepted set of truths.

The proof of Pythagoras' theorem has a structure that can be clearly defined in terms of a sequence of propositions culminating in a conclusion. It forms what is called an argument, which consists of a collection of accepted truths arranged in such a manner that the conclusion is demonstrated to be a necessary consequence of the reasoning. This notion of an argument is crucial to the proof and, indeed, before we can study the proofs of theorems in any detail we need to understand the structure of logical arguments.

Arguments

The everyday use of the word argument refers to a dispute that more often than not is caused by two or more people coming to different conclusions about the same shared information. In reality such arguments have more to do with rhetoric than logical reasoning.

In logic the word argument is taken to mean a sequence of statements called premises followed by a single statement called the conclusion. The purpose of the argument is to derive or to infer a valid conclusion from the given set of premises, and if this is achieved then the argument is said to be a valid argument. For example, the following sequence of statements form a valid argument:

If I select the colour red then I select a colour of the rainbow

I select the colour red

therefore

I select a colour of the rainbow

The form of this argument consists of two premises ('If I select the colour red then I select a colour of the rainbow' and 'I select the colour red') and a single conclusion ('I select a colour of the rainbow').

An argument of this form – a three-line argument consisting of two given premises and a single conclusion – is called a **syllogism**. This particular syllogism is of a type referred to as *modus ponens*, which is Latin for **method of affirming**. Another syllogism of a type referred to as *modus tollens* – or **method of denying** – is exemplified by

If I select the colour red then I select a colour of the rainbow

I did not select a colour of the rainbow

therefore

I did not select the colour red

You may notice that this argument is related to the contrapositive of the first argument. We shall see this connection more clearly if we turn to a more general notation.

Assign the following propositions to the variables p and q where

p = I select the colour red

q = I select a colour of the rainbow

Modus ponens can then be written **structurally** as

$$\frac{p \rightarrow q, p}{q}$$

to be read as

given that p implies q and given p then we can conclude q

or

from $p \rightarrow q$ and p we can derive q

The structure of the form of this argument consists of all the given premises being written above the line and the derived conclusion being written below the line.

Modus tollens can be written structurally as:

$$\frac{p \rightarrow q, \neg q}{\neg p}$$

to be read as

given that p implies q and given not q then we can conclude not p

or

from $p \rightarrow q$ and $\neg q$ we can derive $\neg p$

The fact that these are two valid arguments can be further seen if we couple the premises with the operation AND, IMPLY the conclusion and complete the truth table. This means that *modus ponens* becomes

$$((p \rightarrow q) \wedge p) \rightarrow q$$

The truth table for this compound proposition is then

p	**q**	**p**	$\rightarrow$	**q**	$\wedge$	**p**	$\rightarrow$	**q**
0	0	0	1	0	0	0	1	0
0	1	0	1	1	0	0	1	1
1	0	1	0	0	0	1	1	0
1	1	1	1	1	1	1	**1**	1
							*	

The final implication (*) demonstrates that the compound proposition is a tautology, which indicates that the form or structure of the argument is valid.

For *modus tollens* we have

$$((p \rightarrow q) \wedge \neg q) \rightarrow \neg p$$

which has the following truth table:

p	**q**	**p**	$\rightarrow$	**q**	$\wedge$	**¬q**	$\rightarrow$	**¬p**
0	0	0	1	0	1	1	1	1
0	1	0	1	1	0	0	1	1
1	0	1	0	0	0	1	1	0
1	1	1	1	1	0	0	1	0
							*	

Again, the final implication (*) is a tautology, thereby demonstrating that the structure of the argument is valid.

The fact that both expressions are tautologies indicates that they are true regardless of the specific contents of the propositions, hence demonstrating that they are syntactically valid.

Conclusion: Every argument consists of a collection of premises and a final conclusion. The process of proving arguments to be valid is one of demonstrating that any stated assertion within the argument is consistent with everything that has been said before and that the conclusion is a necessary consequence of the mathematical framework within which the proof is derived. Arguments can be proved to be valid by constructing the compound proposition formed from the conjunction of the argument's premises implying the argument conclusion. If the resulting compound proposition is a tautology then the argument is valid.

Worked example 6.1

Demonstrate the validity of each of the following arguments by constructing the appropriate tautology:

(a) Conjunctive simplification

$$\frac{p \wedge q}{p} \quad \text{and} \quad \frac{p \wedge q}{q}$$

(b) Disjunctive syllogism

$$\frac{p \vee q, \neg q}{p}$$

(c) Division into cases

$$\frac{p \vee q, p \to r, q \to r}{r}$$

Solution:

(a)

$$\frac{p \wedge q}{p} \quad \text{and} \quad \frac{p \wedge q}{q}$$

The arguments give rise to the following implications:

p	q	(p	∧	q)	→	p	and	(p	∧	q)	→	q
0	0	0	0	0	1	0		0	0	0	1	0
0	1	0	0	1	1	0		0	0	1	1	1
1	0	1	0	0	1	1		1	0	0	1	0
1	1	1	1	1	1	1		1	1	1	1	1
					*						*	

The existence of the tautologies ensures the validity of the arguments.

(b)

$$\frac{p \vee q, \neg q}{p} \quad \text{and} \quad \frac{p \vee q, \neg p}{q}$$

The arguments give rise to the following implications:

p	q	[(p	∨	q)	∧	¬q]	→	p	and	[(p	∨	q)	∧	¬p]	→	q
0	0	0	0	0	0	1	1	0		0	0	0	0	1	1	0
0	1	0	1	1	0	0	1	0		0	1	1	1	1	1	1
1	0	1	1	0	1	1	1	1		1	1	0	0	0	1	0
1	1	1	1	1	0	0	1	1		1	1	1	0	0	1	1
							*								*	

The existence of the tautologies ensures the validity of the arguments.

(c)

$$\frac{p \vee q, p \to r, q \to r}{r}$$

The argument gives rise to the following implication:

p	q	r	{[(p	∨	q)	∧	(p	→	r)]	∧	(q	→	r)}	→	r
0	0	0	0	0	0	0	0	1	0	0	0	1	0	1	0
0	0	1	0	0	0	0	0	1	1	0	0	1	1	1	1
0	1	0	0	1	1	1	0	1	0	0	1	0	0	1	0
0	1	1	0	1	1	1	0	1	1	1	1	1	1	1	1
1	0	0	1	1	0	0	1	0	0	0	0	1	0	1	0
1	0	1	1	1	0	1	1	1	1	1	0	1	1	1	1
1	1	0	1	1	1	0	1	0	0	0	1	0	0	1	0
1	1	1	1	1	1	1	1	1	1	1	1	1	1	1	1
														*	

The existence of the tautologies ensures the validity of the arguments.

Worked example 6.2

Demonstrate that each of the following arguments is invalid:

(a) Inverse error

$$\frac{p \rightarrow q, \neg p}{\neg q}$$

(b) $$\frac{p \rightarrow q, q \rightarrow p}{p \vee q}$$

Solution:

(a) $$\frac{p \rightarrow q, \neg p}{\neg q}$$

This argument gives rise to the following implication:

p	q	[(p	→	q)	∧	¬p]	→	¬q
0	0	0	1	0	1	1	1	1
0	1	0	1	1	1	1	0	0
1	0	1	0	0	0	0	1	1
1	1	1	1	1	1	0	1	0
							*	

This is not a tautology so the argument is invalid.

(b) $$\frac{p \rightarrow q, q \rightarrow p}{p \vee q}$$

This argument gives rise to the following implication:

p	q	[(p	→	q)	∧	(q	→	p)]	→	p	∨	q
0	0	0	1	0	1	0	1	0	0	0	0	0
0	1	0	1	1	0	1	0	0	1	0	1	1
1	0	1	0	0	0	0	1	1	1	1	1	0
1	1	1	1	1	1	1	1	1	1	1	1	1
									*			

This is not a tautology so the argument is invalid.

Exercises

6.1 Demonstrate the validity of each of the following arguments:
  (a) *Modus tollens*

$$\frac{p \rightarrow q, \neg q}{\neg p}$$

  (b) Disjunctive addition

$$\frac{p}{p \vee q} \quad \text{and} \quad \frac{q}{p \vee q}$$

  (c) Hypothetical syllogism

$$\frac{p \rightarrow q, q \rightarrow r}{p \rightarrow r}$$

6.2 Demonstrate that each of the following arguments is invalid:
  (a) Converse error

$$\frac{p \rightarrow q, q}{p}$$

  (b) $$\frac{(p \wedge q) \rightarrow r, \neg q}{p \rightarrow r}$$

Completeness and consistency

When we declared the first argument form, the *modus ponens* argument, to be valid we did so initially by appealing to the actual meaning of what was being declared. That is we appealed to the **semantics** of the argument to demonstrate its validity. When we declared the formal form of the argument to be true we appealed to the **syntax** of the argument – the structure or form of the argument. When we constructed the implication and looked at the truth table we were at a sort of halfway house. We were using a semantic argument in terms of truth values to demonstrate that the syntax of the argument was correct.

Being able to demonstrate the validity of an argument by consideration of its syntax (its structure) alone demonstrates the **completeness** of the theory of propositions. Being also able to demonstrate the validity of the same argument by appeal to its semantics (its meaning) demonstrates the **consistency** of the theory of propositions. Furthermore, any complete argument is consistent and any consistent argument is complete. It is this very completeness and consistency that enables us to build computers that obey all the laws of logic that we have here espoused.

While you may think that we are making rather heavy weather about the distinctions between syntax and semantics, consider the first four lines of a poem called Jabberwocky by the logician the Reverend Charles Dodgson, known to most as Lewis Carroll:

> 'Twas brillig and the slithy toves
> Did gyre and gimble in the wabe
> All mimsy were the borogroves
> And the mome raths outgrabe.

Syntactic perfection but semantic nonsense. Better still is the following penned by Luis d'Antin van Rooten:

> Lille beau pipe
> Ocelot serre chypre
> En douzaine aux verres tuf indemne
> Livre de melons un dé huile qu'aux mômes
> Eau à guigne d'air telle baie indemne.

To appreciate the effect of this little poem you must read it out loud with a French accent.

Just as we have stated that syntactic proofs are derived within a given context, so it must be realized that semantics must also be considered within a specified context. What makes the Jabberwocky so tantalizing is the presence of evocative yet non-contextual words such as 'brillig', 'gimble' and 'mimsy' combined with recognized words such as 'and', 'the' and 'all' coupled together within a recognized syntactic framework. Indeed, it is just this structure that has inspired people to guess at the meanings of the various nonsense words in the passage and to explain its meaning as a whole – an activity which misses the point of the passage completely. Van Rooten's offering, on the other hand, is surreal: a recognizable syntactic rhythm and a French vocabulary sufficiently well defined to induce the feeling of poetry which, when read out loud, renders a well-known English semantic and syntactic structure.

Theorems

A theorem in logic is a statement in the form of an expression involving propositions, variables, quantifiers and the logical connectives. To prove a theorem we have to demonstrate that the theorem can be derived from what we know to be true. What we know to be true in the theory of propositional and predicate calculus is contained within the rules of logic on pp. 46ff. Indeed, derivations can be and are constructed using these very rules.

In Part Two of this book we were first introduced to the idea of proof when we studied proof by induction. This is a very powerful method of proving appropriate statements and has the distinct advantage that it is relatively easy to see where you are in the proof and where you need to go. In addition, the method used can be generally applied quite systematically to countless situations. On the other hand, when we proved Pythagoras' theorem, although we used a well-established technique of arranging premises in a logical order, the method of constructing the proof was special to that particular theorem.

In either case, proof by induction or the proof of Pythagoras' theorem, we made use of a context with which the theorem was proved. This context of accepted truths consists of a collection of axioms.

Axioms

An axiom is a **defined** truth or a **self-evident** truth. For example, the statement

If you walk in the rain then you are going to get wet

is a self-evident truth. We do not need to walk in the rain to prove that it is true, we accept the fact. On the other hand, the statement

$1 + 1 = 2$

is a defined truth. It may seem as though it is self-evident, but that is only because you have become so immersed within the culture that accepts it that it appears to be self-evident. In fact it is a definition – it cannot be proved because it is fundamental to arithmetic and does not rest upon earlier definitions.

The list of laws governing the algebra of propositions that we met earlier in Chapter 2 are axioms – they are defined truths. That they can be demonstrated to be semantically reasonable by applying them to the algebra of sets or to the logic of English sentences does not make them self-evident – they only appear to be so. These axioms provide a context within which we can prove our statements in the propositional calculus to be correct. We have seen this in Chapter 2 when we proved that

$[(p \wedge \neg q) \vee (\neg p \wedge r)] \vee (p \wedge q) \equiv (p \vee r)$

Notice that we stated that the axioms of the propositional calculus provided **a** context and not **the** context. There are other, alternative, sets of axioms that perform the function of providing a context equally well – if not better in certain circumstances. It is horses for courses. Recall that we defined the natural numbers in terms of the empty set and that from this we were able to consistently develop the arithmetic of the natural numbers. Here, instead of using the arithmetic axioms as given, we derive theorems about the natural numbers using the axioms of sets as our context – in this approach sets become more fundamental to mathematics than the natural numbers.

The ubiquity of the computer means that the correctness of the software developed is critical. One only has to think of particular examples for this statement to be seen to be correct: fly-by-wire aircraft controlled by a pilot using a computer; a nuclear power station whose entire operation is computer controlled; a rail network in which scheduling and traffic flow are computer controlled. There are numerous instances such as these where a software failure could have catastrophic consequences. At one time all software was tested by using trial data, but this method is not adequate for safety-critical situations. The trial data cannot exhaust all possibilities of data input nor can it guarantee to have covered all possible permutations of software operation.

What is required is that we prove that our software constructions are correct using well-defined mathematical structures. However, programs are now so huge that to prove them manually becomes a virtually impossible task – impossible in the sense of the time and the money available. To overcome this we need (and have) programs that can be used to automate the proving process.

If we are to automate proofs we need a context – a set of axioms – and a standardized method of applying them: standardized in much the same way that inductive proof is standardized. We could use the axioms of the propositional calculus as indeed we have done in the earlier parts of this book, but in the event this proves to be a most inefficient method. An alternative, more efficient, method is provided by using a set of axioms known as the Gentzen rules of natural deduction.

The Gentzen rules of natural deduction
The two syllogisms of *modus ponens* and *modus tollens* derive from the Middle Ages when the students of antiquity were taught the seven scholastic disciplines that constituted the trivium and the quadrivium. Of the seven, the three disciplines of grammar, rhetoric and logic were contained in the trivium, and this is where the Latin names originate.

There are other valid argument forms, each one of which can be shown to be both semantically and syntactically correct. However, we shall not consider argument forms *per se* but instead look at the theory of propositions as an algebra of symbols upon which are imposed a set of rules or axioms.

We need to be able to derive conclusions from premises, and it is this procedure on which we shall now focus our attention. Our starting point will be the rules – the Gentzen rules of natural deduction – which deal with the ability to eliminate or introduce the five logical connectives of $\wedge$, $\vee$, $\neg$, $\rightarrow$ and $\leftrightarrow$ within expressions involving propositions.

$\wedge$ Elimination

$$\frac{p \wedge q}{p} \quad \text{and} \quad \frac{p \wedge q}{q}$$

Here we see that given p AND q then we can derive p or we can derive q. The truth of these two rules is self-evident but it can be demonstrated by constructing the truth tables of

$$p \land q \to p \text{ and } p \land q \to q$$

both of which are tautologies.

$\land$ Introduction

$$\frac{p, q}{p \land q} \text{ and } \frac{p, q}{q \land p}$$

Given p and given q we can derive p AND q or we can derive q AND p.

Notice that from these two rules we can deduce the commutativity of the AND operator, whereas if we were to use the rules given in Chapter 2 then the commutativity of the AND operator would have been one of the rules itself.

$\lor$ Introduction

$$\frac{p}{p \lor q} \text{ and } \frac{p}{q \lor p}$$

Given p we can derive p OR q or we can derive q OR p.

As an example of the use of these three rules in a derivation we can show that we can derive $p \lor q$ from $p \land q$:

1. $p \land q$ a given premise
2. p derived by using the $\land$ elimination rule
3. $p \lor q$ derived by using the $\lor$ introduction rule

Consequently, we can derive $p \lor q$ from $p \land q$. To represent this fact we use the symbol

$$\vdash$$

called the **syntactic turnstile**.

$$p \land q \vdash p \lor q$$

means that $p \lor q$ can be derived syntactically (that is just by using the rules of natural deduction) from $p \land q$.

We can also demonstrate that this derivation is possible by constructing the truth table of

$$p \land q \to p \lor q$$

Constructing a truth table to demonstrate the validity of a compound proposition is a semantic demonstration because assigning truth values to a proposition is to give it meaning. The truth table is as follows:

p	q	p	∧	q	→	p	∨	q
0	0	0	0	0	**1**	0	0	0
0	1	0	0	1	**1**	0	1	1
1	0	1	0	0	**1**	1	1	0
1	1	1	1	1	**1**	1	1	1
					*			

The fact that this implication is a tautology demonstrates that $p \vee q$ can also be semantically derived from $p \wedge q$. A semantic derivation is indicated by the **semantic turnstile**

$$\vDash$$

so that

$$p \wedge q \vDash p \vee q$$

The fact that we have demonstrated this both syntactically and semantically demonstrates completeness and consistency.

Conclusion: Arguments in the propositional calculus are proved valid by appealing either to their meaning (semantics) or to their structure (syntax). In either case such proofs are constructed within a context provided by a set of self-evident truths called axioms. The Gentzen rules of natural deduction form a set of axioms that can be used as an alternative to the axioms contained within the algebra of the propositional calculus.

Worked example 6.3

Demonstrate that
(a) $p \wedge q \vDash q \wedge p$
(b) $p \wedge (q \wedge r) \vDash (p \wedge q) \wedge r$
(c) $p, q \wedge r \vDash p \wedge r$

Solution:
(a) This is proved by showing that $p \wedge q \rightarrow q \wedge p$ is a tautology:

p	q	(p	∧	q)	→	(q	∧	p)
0	0	0	0	0	**1**	0	0	0
0	1	0	0	1	**1**	1	0	0
1	0	1	0	0	**1**	0	0	1
1	1	1	1	1	**1**	1	0	1
					*			

Hence $p \wedge q \vDash q \wedge p$.

(b) This is proved by showing that p ∧ (q ∧ r) → (p ∧ q) ∧ r is a tautology:

p	q	r	p	∧	(q	∧	r)	→	(p	∧	q)	∧	r
0	0	0	0	0	0	0	0	1	0	0	0	0	0
0	0	1	0	0	0	0	1	1	0	0	0	0	1
0	1	0	0	0	1	0	0	1	0	0	1	0	0
0	1	1	0	0	1	1	1	1	0	0	1	0	1
1	0	0	1	0	0	0	0	1	1	0	0	0	0
1	0	1	1	0	0	0	1	1	1	0	0	0	1
1	1	0	1	0	1	0	0	1	1	1	1	0	0
1	1	1	1	1	1	1	1	1	1	1	1	1	1
								*					

Hence p ∧ (q ∧ r) ⊨ (p ∧ q) ∧ r.

(c) This is different from the previous two examples because we have more than one premise. The principle of proving the semantic deduction is, however, the same. We prove the semantic deduction by showing that p ∧ (q ∧ r) → (p ∧ r) is a tautology:

p	q	r	p	∧	(q	∧	r)	→	(p	∧	r)
0	0	0	0	0	0	0	0	1	0	0	0
0	0	1	0	0	0	0	1	1	0	0	1
0	1	0	0	0	1	0	0	1	0	0	0
0	1	1	0	0	1	1	1	1	0	0	1
1	0	0	1	0	0	0	0	1	1	0	0
1	0	1	1	0	0	0	1	1	1	0	1
1	1	0	1	0	1	0	0	1	1	1	0
1	1	1	1	1	1	1	1	1	1	1	1
								*			

Hence p, q ∧ r ⊨ p ∧ r.

Worked example 6.4

Demonstrate that
(a) p ∧ q ⊢ q ∧ p
(b) p ∧ (q ∧ r) ⊢ (p ∧ q) ∧ r
(c) p, q ∧ r ⊢ p ∧ r

Solution:
(a) 1. p ∧ q   given premise
     2. q   1 and ∧ elimination
     3. p   1 and ∧ elimination
     4. q ∧ p   2, 3 and ∧ introduction
     Therefore p ∧ q ⊢ q ∧ p.

(b) 1. p ∧ (q ∧ r)   given premise
     2. p   1 and ∧ elimination

3. q ∧ r   1 and ∧ elimination
4. q   3 and ∧ elimination
5. p ∧ q   2, 4 and ∧ introduction
6. r   3 and ∧ elimination
7. (p ∧ q) ∧ r   5, 6 and ∧ introduction
Therefore p ∧ (q ∧ r) ⊢ (p ∧ q) ∧ r.

(c) 1. p   given premise
     2. q ∧ r   given premise
     3. r   2 and ∧ elimination
     4. p ∧ r   1, 3 and ∧ introduction
Therefore p, q ∧ r ⊢ p ∧ r.

Exercises

6.3 Demonstrate that
(a) p ∧ (p ∨ q) ⊨ (p ∨ q) ∧ p
(b) p ∧ (q ∧ [r ∧ s]) ⊨ (p ∧ q) ∧ (r ∧ s)
(c) p, q ∧ r ⊨ (p ∧ q) ∧ (p ∧ r)

6.4 Demonstrate that
(a) p ∧ (p ∨ q) ⊢ (p ∨ q) ∧ p
(b) p ∧ (q ∧ [r ∧ s]) ⊢ (p ∧ q) ∧ (r ∧ s)
(c) p, q ∧ r ⊢ (p ∧ q) ∧ (p ∧ r)

The role of assumptions

Assumptions
The proofs that we have considered so far have relied on given premises. Many proofs, however, rely not only on given premises but also on assumptions. As a particular example we shall consider the proof of the theorem that states that

√2 is not a rational number

You will remember that a **rational** number is any number that can be expressed as a **ratio** of integers. Any number that cannot be so expressed is called an **irrational** number, and the theorem that we shall now derive states that √2 is an irrational number.

Theorem

√2 is an irrational number

Proof
The essence of this proof lies in the use of an assumption.

1. *Assumption.* Assume that $\sqrt{2}$ is a rational number of the form m/n. That is

    $\sqrt{2} = m/n$

    Here we have introduced an **assumption** that the theorem is **not true**.

2. *Premise.* The ratio m/n is in its lowest form (that is all common factors have been cancelled out).

3. *Derivation.* Since $\sqrt{2} = m/n$, then by squaring both sides we see that

    $2 = m^2/n^2$

    that is

    $m^2 = 2n^2$

    Therefore m^2 is an even number (it can be divided by 2 to give n^2).

4. *Derivation.* m is even (because m^2 is even). Write m explicitly as an even number:

    $m = 2p$

    where p is an integer.

5. *Derivation.* Therefore, by substitution

    $m^2 = (2p)^2 = 4p^2 = 2n^2$

    so that

    $n^2 = 2p^2$

    Hence n^2 is an even number.

6. *Derivation.* Hence n is an even number. Write n explicitly as an even number:

    $n = 2q$

    where q is an integer.

7. *Derivation.* Therefore

    $m/n = 2p/2q = p/q$

8. *Derivation.* Therefore m/n is not in its lowest form: both m and n have a factor 2 in common. This statement **contradicts** the premise given in 2 that they did not have a factor in common.

9. *Conclusion.* Based upon the initial assumption, the proof has produced a contradiction in that a given premise is shown to be inconsistent with the assumption. This shows that the assumption is incorrect thus proving that

    $\sqrt{2}$ is an irrational number

This derivation, known as *reductio ad absurdum* (literally reduction to absurdity or proof by contradiction), has made use of the requirement that any assumptions introduced into a proof must be consistent with the given premises if the assumption is to be accepted as valid. Premises are inviolable because they are given facts and constitute the starting point from which the proof is derived. Assumptions can be introduced into a proof, but they are only valid insofar as they are consistent with the given premises. Furthermore, the introduction of an assumption is not a hit-and-miss affair; an assumption is introduced into an argument specifically to permit a rule of logic to be initiated. When that rule has been initiated the assumption has run its course and is discharged, playing no further role in the derivation. For example, consider the proof that

$$(p \wedge q) \vee (p \wedge \neg q) \vdash p$$

Notice that the premise $(p \wedge q) \vee (p \wedge \neg q)$ does not state that any one of the premise components p, q, $\neg q$, $(p \wedge q)$ or $(p \wedge \neg q)$ is given. To use any one of the components we must **assume** it – we must introduce it as an assumption. To prove this syntactic derivation we need to make use of a further rule of natural deduction, namely the rule for OR elimination

∨ **Elimination**

$$\frac{[P, a \vdash p; \ P, b \vdash p; \ a \vee b]}{p}$$

The OR elimination rule requires a little explanation. The rule states that if a collection of premises represented by P, coupled with the assumption a, can be used to derive p, and the same collection of premises P coupled with the assumption b can also be used to derive p and that a ∨ b is a given premise, then we can derive p.

We can now proceed to the proof of $(p \wedge q) \vee (p \wedge \neg q) \vdash p$ noting that, while we have the given premise $(p \wedge q) \vee (p \wedge \neg q)$, we do not have either component as a premise. Consequently, in order to use these components we shall need to introduce them as assumptions:

1.	$(p \wedge q) \vee (p \wedge \neg q)$	given premise	this is P
2.	$[p \wedge q$	assumption 1	this is a

Notice that we have opened a bracket here. At this stage of the proof we need to assume $p \wedge q$ in order to progress. We indicate the introduction of an assumption by the inclusion of the square bracket to give a visual indication that an assumption has been introduced.

3. p 2 and ∧ elimination

Here we see that we have derived p under the assumption introduced at 2, that is we have completed the derivation

$$P, a \vdash p$$

which is the first part of the ∨ elimination rule.

4.   [p ∧ ¬q   assumption 2   this is b

Notice that we have opened a second bracket here. Here we have assumed the second component of the initial premise.

5.   p]]   4 and ∧ elimination

Notice that we have closed both brackets here.
   Here we see that we have derived p under the assumption introduced at 4, that is we have completed the derivation

   P, b ⊢ p

which is the second part of the ∨ elimination rule.
   Having completed the requirements of the ∨ elimination rule we can now invoke the rule.

6.   p   3, 5 and ∨ elimination

Notice the two square brackets in 5. A closing square bracket indicates that an assumption has been discharged by the application of the rule for which the assumption was required; in this case we have discharged the two assumptions simultaneously by application of the ∨ elimination rule. Therefore (p ∧ q) ∨ (p ∧ ¬q) ⊢ p.
   We are now in a position to list the 10 Gentzen rules.

The Gentzen rules of natural deduction
In what follows p and q are propositions, P represents a collection of premises and a and b are assumptions.

∧ Introduction

$$\frac{p, q}{p \wedge q} \quad \textbf{and} \quad \frac{p, q}{q \wedge p}$$

∧ Elimination

$$\frac{p \wedge q}{p} \quad \textbf{and} \quad \frac{p \wedge q}{q}$$

∨ Introduction

$$\frac{p}{p \vee q} \quad \textbf{and} \quad \frac{q}{p \vee q}$$

∨ Elimination

$$\frac{[P, a \vdash p, P, b \vdash p, a \vee b]}{p}$$

¬ Introduction

$$\frac{[P, a \vdash p, P, a \vdash \neg p]}{\neg a}$$

¬ Elimination

$$\frac{\neg \neg p}{p}$$

→ Introduction

$$\frac{[P, a \vdash p]}{a \rightarrow p}$$

→ Elimination

$$\frac{p, p \rightarrow q}{q}$$

↔ **Introduction**   ↔ **Elimination**

$$\frac{p \rightarrow q,\, q \rightarrow p}{p \leftrightarrow q}$$ $$\frac{p \leftrightarrow q}{p \rightarrow q} \quad \text{and} \quad \frac{p \leftrightarrow q}{q \rightarrow p}$$

Conclusion: An assumption is a proposition introduced into an argument to facilitate its derivation. Every assumption is introduced to permit one of the rules of the logic to be initiated. When that rule has been initiated the assumption is discharged and plays no further role in the derivation. The Gentzen rules of natural deduction are rules of logic that enable the introduction and elimination of the operators of the propositional calculus.

Worked example 6.5

Demonstrate that

(a) $p \vee (p \wedge q) \models p$
(b) $p \wedge (q \vee r) \models (p \wedge q) \vee (p \wedge r)$
(c) $(p \rightarrow q) \wedge p \models q$
(d) $p \leftrightarrow q \models (p \rightarrow q) \wedge (q \rightarrow p)$

Solution:

(a) This is proved by showing that $p \vee (p \wedge q) \rightarrow p$ is a tautology:

p	q	p	∨	(p	∧	q)	→	p
0	0	0	0	0	0	0	1	0
0	1	0	0	0	0	1	1	0
1	0	1	1	1	0	0	1	1
1	1	1	1	1	1	1	1	1

Hence $p \vee (p \wedge q) \models p$.

(b) This is proved by showing that $p \wedge (q \vee r) \rightarrow (p \wedge q) \vee (p \wedge r)$ is a tautology.

p	q	r	p	∧	(q	∨	r)	→	(p	∧	q)	∨	(p	∧	r)
0	0	0	0	0	0	0	0	1	0	0	0	0	0	0	0
0	0	1	0	0	0	1	1	1	0	0	0	0	0	0	1
0	1	0	0	0	1	1	0	1	0	0	1	0	0	0	0
0	1	1	0	0	1	1	1	1	0	0	1	0	0	0	1
1	0	0	1	0	0	0	0	1	1	0	0	0	1	0	0
1	0	1	1	1	0	1	1	1	1	0	0	1	1	1	1
1	1	0	1	1	1	1	0	1	1	1	1	1	1	0	0
1	1	1	1	1	1	1	1	1	1	1	1	1	1	1	1

Hence $p \wedge (q \vee r) \models (p \wedge q) \vee (p \wedge r)$.

(c) This is proved by showing that $[(p \rightarrow q) \wedge p] \rightarrow q$ is a tautology:

p	q	[(p	→	q)	∧	p]	→	q
0	0	0	1	0	0	0	**1**	0
0	1	0	1	1	0	0	**1**	1
1	0	1	0	0	0	1	**1**	0
1	1	1	1	1	1	1	**1**	1

Hence $(p \rightarrow q) \wedge p \vDash q$.

(d) This is proved by showing that $(p \leftrightarrow q) \rightarrow [(p \rightarrow q) \wedge (q \rightarrow p)]$ is a tautology:

p	q	(p	↔	q)	→	[(p	→	q)	∧	(q	→	p)]
0	0	0	1	0	**1**	0	1	0	1	0	1	0
0	1	0	0	1	**1**	0	1	1	0	1	0	0
1	0	1	0	0	**1**	1	0	0	0	0	1	1
1	1	1	1	1	**1**	1	1	1	1	1	1	1

Hence $p \leftrightarrow q \vDash (p \rightarrow q) \wedge (q \rightarrow p)$

Worked example 6.6

Demonstrate that
(a) $p \vee (p \wedge q) \vdash p$
(b) $p \wedge (q \vee r) \vdash (p \wedge q) \vee (p \wedge r)$
(c) $(p \rightarrow q) \wedge p \vdash q$
(d) $p \leftrightarrow q \vdash (p \rightarrow q) \wedge (q \rightarrow p)$

Solution:

(a) 1. $p \vee (p \wedge q)$ given premise
    2. $[p \wedge q$ assumption
    3. $p]$ 2 and ∧ elimination
    4. p 1, 2, 3 and ∨ elimination
Therefore $p \vee (p \wedge q) \vdash p$.

(b) 1. $p \wedge (q \vee r)$ given premise
    2. p ∧ elimination
    3. $q \vee r$ ∧ elimination
    4. $[q$ assumption
    5. $p \wedge q$ ∧ introduction
    6. $[r$ assumption
    7. $p \wedge r]]$ ∧ introduction
    8. $(p \wedge q) \vee (p \wedge r)$ ∨ introduction

(c) 1. $(p \rightarrow q) \wedge p$ given premise
    2. p ∧ elimination
    3. $p \rightarrow q$ ∧ elimination
    4. q → elimination

(d) 1. $p \leftrightarrow q$ given premise

2. $p \rightarrow q$ $\leftrightarrow$ elimination
3. $q \rightarrow p$ $\leftrightarrow$ elimination
4. $(p \rightarrow q) \wedge (q \rightarrow p)$ $\wedge$ introduction

---

Exercises

6.5 Demonstrate that
   (a) $p \vee q \vDash q \vee p$
   (b) $p \wedge (p \vee q) \vDash p$
   (c) $(p \rightarrow q) \wedge (q \rightarrow r) \vDash (p \rightarrow r)$
   (d) $p \vee (q \wedge r) \vDash (p \vee q) \wedge (p \vee r)$

6.6 Demonstrate that
   (a) $p \vee q \vdash q \vee p$
   (b) $p \wedge (p \vee q) \vdash p$
   (c) $(p \rightarrow q) \wedge (q \rightarrow r) \vdash (p \rightarrow r)$
   (d) $p \vee (q \wedge r) \vdash (p \vee q) \wedge (p \vee r)$

Gentzen rules involving quantifiers
In addition to the 10 rules for the introduction and elimination of the five logical operators there are four further rules dealing with the introduction and the elimination of the logical quantifiers.

Universality
The two rules for the introduction and the elimination of the universal quantifier are as follows.

$\forall$ *Elimination*

$$\frac{\forall x \in D, P(x)}{P(\alpha)} \quad \alpha \text{ arbitrary}$$

This rule states that if we are given premise $P(x)$ for all values of $x \in D$ then we can derive $P(x)$ for any specific value of $x \in D$, in particular $x = \alpha$ where α is arbitrary.

$\forall$ *Introduction*

$$\frac{P(\alpha)}{\forall x \in D, P(x)} \quad \alpha \text{ arbitrary}$$

This rule states that if we are given premise $P(\alpha)$ for any value of $\alpha \in D$ where α is arbitrary then we can derive $P(x)$ for all values of $x \in D$.

Existence
The two rules for the introduction and the elimination of the existential quantifier are as follows.

∃ Introduction

$$\frac{P(\beta)}{\exists x \in D, P(x)}$$

This rule states that if we are given premise $P(\beta)$ for at least one specific value of $\beta \in D$ then we can derive the existential statement $\exists x \in D, P(x)$.

∃ Elimination

$$\frac{\exists x \in D, P(x), \forall x \in D, [P(x) \rightarrow Q]}{Q}$$

This rule states that if we are given the existential premise $\exists x \in D, P(x)$, which states the $P(x)$ is true for at least one element of D, and the universal premise $\forall x \in D, [P(x) \rightarrow Q]$, which states that $P(x)$ implies Q for all elements of D, then we can derive Q.

Worked example 6.7

Prove the following derivations:
(a) $\forall x \in D, [P(x) \rightarrow Q(x)] \land \neg Q(\alpha) \vdash \neg P(\alpha)$
(b) $P(\alpha), \forall x \in D, \{P(x) \rightarrow Q(x)\} \vdash Q(\alpha)$
(c) If n^2 is an even integer then n is an even integer

Solution:
(a) 1. $\forall x \in D, [P(x) \rightarrow Q(x)] \land \neg Q(\alpha)$ given premise
     2. $\forall x \in D, [P(x) \rightarrow Q(x)]$ ∧ elimination
     3. $P(\alpha) \rightarrow Q(\alpha)$ $(\alpha \in D)$ ∀ elimination
     4. $\neg Q(\alpha)$ ∧ elimination
     5. $[P(\alpha)$ assumption
     6. $Q(\alpha)]$ 3, 5 and → elimination
     7. $\neg P(\alpha)$ 4, 5, 6 and ¬ introduction
(b) 1. $\forall x \in D, \{P(x) \rightarrow Q(x)\}$ given premise
     2. $P(\alpha)$ premise
     3. $P(\alpha) \rightarrow Q(\alpha)$ 1 and ∀ elimination
     4. $Q(\alpha)$ 2, 3 and → elimination
(c) Define the predicate EVEN(n) = 'n is an even integer'. We are required to prove that EVEN$(n) \rightarrow$ EVEN(n^2) for all $n \in Z$, that is

$\vdash \forall n \in Z, [\text{EVEN}(n) \rightarrow \text{EVEN}(n^2)]$

Notice that we do not have any premises. We must begin with an assumption – we assume that we have an integer which is even:

1. [EVEN (n) assumption
2. $\exists p \in Z, n = 2p$ $\exists$ introduction
3. $n = 2m$ $\exists$ elimination
4. $n^2 = 4m^2$
5. $n^2 = 2(2m^2)$
6. $\exists q \in Z, n^2 = 2q$ $\exists$ introduction
7. EVEN(n^2) $\exists$ elimination
8. EVEN$(n) \rightarrow$ EVEN$(n^2)]$ $\rightarrow$ introduction
9. $\forall n \in Z, \{\text{EVEN}(n) \rightarrow \text{EVEN}(n^2)\}$ $\forall$ introduction

---

Exercise

6.7 Prove the following derivations:

(a) $\forall x \in D, [P(x) \rightarrow Q(x)] \wedge P(\alpha) \vdash Q(\alpha)$

(b) $\forall x \in D, P(x) \vdash \exists y \in D, P(x)$

(c) If n is an odd integer then n^2 is an odd integer

Part Four

Relations and Functions

We are all familiar with the idea of a relation as being someone to whom we are related, but to use such an idea as the definition of a relation is too self-referential for it to be worthwhile. All relations are subject to a relationship, and relationships abound, not least in the world of computing. Every piece of software bears a relationship to the type of machine on which it can be used, every central processing unit bears a relationship to the type of operating system that it will support and each application package bears a relationship to the uses to which it will be put. Such statements, however, are too vague to be meaningful and useful. To make use of these ideas we need to formalize our concept of relationship and relation and in this part that is what we do.

Another concept that is familiar is that of a function but, like the word relation, the word function can be used in a number of different senses. For example, an algorithm is a sequence of operations designed to perform some function and, as we were all told in High School, y is a function of x. In the final chapter we consider functions and their graphs and we shall discover a close connection between the concepts of a relation and a function.

Chapter 7

Relations

OBJECTIVES

When you have completed this chapter you will be able to:

☐ explain the distinction between a relationship and a relation;

☐ construct the digraph of a binary relation;

☐ identify an order relation and distinguish between total and partial order;

☐ detect the various properties of relations;

☐ identify an equivalence relation;

☐ construct the equivalence classes of an equivalance relation;

☐ use a tree to represent an arithmetic expression;

☐ execute a pre-, post- or in-order search on a tree;

☐ evaluate an arithmetic expression written in either prefix or postfix notation.

Every organization that uses a computer system to maintain its operational and management information does so using a collection of interrelated files called a database. Every file in a database consists of a template called a **record template**, which is used to create individual records of information. Each record contains a collection of predetermined fields into which the record's information is entered. A simple equivalent to a computer file is a physical address book which contains the names, addresses and telephone numbers of your friends and acquaintances. Just as in a computer file, each record in the address book has the same format – one line for name, three lines for address, one line for post code and one line for telephone number.

Using and managing a database requires a detailed knowledge of a mathematical entity called a relation. For example, if you were to open your address book at the page labelled B you would find listed there all your friends and acquaintances whose surname begin with the letter B. You have used the relationship that exists between the letter B on the right-hand side of the page and the collection of names and addresses that the page contains.

Relationships and relations

Relationships
If I were to point to my two labradors and say

Those two dogs are sisters

then I should not only be asserting the fact that the two dogs are related but I should also be describing the type of relationship that existed between them.

Consider the set of ordered pairs of real numbers formed by the cross-product $\mathcal{R} \times \mathcal{R}$:

$\{(x, y): x, y \in \mathcal{R}\}$

and the assertion that expresses a relationship between the elements x and y of each ordered pair:

number x is less than number y

Unlike the previous assertion concerning two specific dogs, this assertion expresses the relationship between any pair of numbers. It does not specify the actual values of the numbers; instead it uses variables to represent them. If we now substitute actual numbers for the variables and test each pair of numbers against the assertion then the assertion will either be true or it will be false. For example

5 is less than 7.85 is a true assertion

and

3.2 is less than -4 is a false assertion

If we collect together all those ordered pairs of numbers whose elements do satisfy the assertion then we form a subset of the original set $\mathcal{R} \times \mathcal{R}$:

$\{(x, y): x, y \in \mathcal{R} \text{ and } x < y\}$

Because the elements of this set satisfy an assertion expressing a relationship we call this set a **relation, defined on the set** $\mathcal{R}$. In particular, we call it a **binary** relation because it is a relation involving a set of ordered **pairs**.

The general form of a binary relation R is:

$$R = \{(x, y): x \in A, y \in B \text{ and } P(x, y)\}$$

where $P(x, y)$ is an open sentence that expresses the relationship between the values of the variables x and y. R is a subset of the cross-product $A \times B$ and consists of those elements of $A \times B$ for which $P(x, y)$ is true.

Ordered *n*-tuples

Relations are not confined to sets of ordered pairs. Relations can also be sets of ordered triples, quadruples or, in general, ordered *n*-tuples. For example, the relation

$$\{(x, y, z): x^2 + y^2 + z^2 = 25, x, y, z \in \mathcal{R}\}$$

is specified by the assertion that the elements of each ordered triple satisfy the equation of a sphere of radius 5, centred on the Cartesian coordinate origin.

Conclusion: A relation is a set of ordered n-tuples where the elements of each ordered n-tuple satisfy an assertion that expresses a relationship between those elements. A relation involving a set of ordered pairs is called a binary relation.

Worked example 7.1

Describe each of the following relations explicitly:
(a) $R = \{(x, y): x, y \in \{0, 1, 2, 3, 4\} \text{ and } x < y\}$
(b) $R = \{(x, y): x \in \{0, 1, 2, 3\}, y \in \{0, 1, 2, 3, 4, 5, 6, 7\} \text{ and } 2x = y\}$
(c) $R = \{(<a>, <b>): <a>, <b> \in \{\text{one, two, three, four}\} \text{ and } <a> \text{ is before } <b> \text{ in the dictionary}\}$
(d) $R = \{(p, q): p \in \{\text{apple, orange, dog, cat}\}, q \in \{\text{fruit, animal, building, colour}\} \text{ and } p \text{ is a } q\}$

Solution:
(a) $\{(0, 1), (0, 2), (0, 3), (0, 4), (1, 2), (1, 3), (1, 4), (2, 3), (2, 4), (3, 4)\}$
(b) $\{(0, 0), (1, 2), (2, 4), (3, 6)\}$
(c) {(four, one), (four, three), (four, two), (one, three), (one, two), (three, two)}
(d) {(apple, fruit), (orange, fruit), (orange, colour), (dog, animal), (cat, animal)}

Worked example 7.2

For each of the following relations, find an assertion that enables the construction of the relation and write down a formal description of the relation:

(a) R = {(0, 0), (1, −1), (2, −2), (3, −3), (−1, 1), (−2, 2), (−3, 3)}
(b) R = {(aaa, aa), (aaa, a), (aba, a), (aba, aa), (aba, aaa), (aa, a)}
(c) R = {(1, 1), (1, −1), (4, 2), (4, −2), (9, 3), (9, −3)}
(d) R = {(aaa, aa), (aaa, a), (aba, a), (aba, aa), (aa, a)}

Solution:
(a) R = {(x, y): x, y ∈ {−3, −2, −1, 0, 1, 2, 3} and x = xy}
(b) R = {(<x>, <y>): <x>, <y> ∈ {a, aa, aaa, aba} and len(<x>) ≥ len(<y>) where <x> ≠ <y>}
(c) R = {(x, y): x ∈ {1, 4, 9}, y ∈ {−3, −2, −1, 0, 1, 2, 3} and x = y^2}
(d) R = {(<x>, <y>): <x>, <y> ∈ {a, aa, aaa, aba} and len(<x>) > len(<y>)}

Exercises

7.1 Describe each of the following relations explicitly:
   (a) R = {(x, y): x, y ∈ {0, 1, 2, 4, 6} and x ≥ y}
   (b) R = {(x, y): x ∈ {−2, −1− 0, 1, 2}, y ∈ {0, 1, 2, 3} and |x| = y}
   (c) R = {(<a>, <b>): <a> ∈ {a, ab, abc, abcd}, <b> ∈ {p, pq, pqr} and the length of <a> is less than the length of <b>}
   (d) R = {(p, q): p, q ∈ {son, daughter, father, mother, paternal grandfather} and p is a descendant of q}

7.2 For each of the following relations, find an assertion that enables the construction of the relation and write down a formal description of the relation:
   (a) R = {(0, 0), (1, 1), (2, 8), (3, 27), (−1, −1), (−2, −8), (−3, −27)}
   (b) R = {(111, 011), (111, 001), (101, 001), (101, 011), (111, 101), (011, 001)}
   (c) R = {(A, B), (A, C), (A, D), (B, C), (B, D), (C, D)}
   (d) R = {(<ten>, <four>), (<ten>, <seven>), (<four>, <seven>)}

Digraphs
A **digraph** is a diagram consisting of **nodes** and **directed edges** that is used to display a binary relation. For example, consider the relation

R = {(x, y): x, y ∈ D and x > y}

where D = {1, 2, 3, 4} is the set on which the relation is defined and where x > y is an open sentence that describes the relationship.

   By substituting pairs of elements of D into the open sentence and testing the truth value of the resulting proposition we find that this relation can be expressed explicitly as

R = {(2, 1), (3, 1), (3, 2), (4, 1), (4, 2), (4, 3)}

A diagram of this relation can be constructed by first representing each element of D by a numbered circle – a **node**. For example, the number 1 is represented by the diagram of Figure 7.1.

Figure 7.1

The inclusion of a specific ordered pair in the relation is indicated by joining the appropriate two elements of D together with an arrow – a **directed edge** – where the arrow points from the first element of the ordered pair to the second. For example, the element (2, 1) of R would be represented by Figure 7.2.

Figure 7.2

This directed graph of the ordered pair is called a **digraph** – a directed graph. The relation R has a digraph that consists of nodes and directed edges representing all the ordered pairs of R (Figure 7.3).

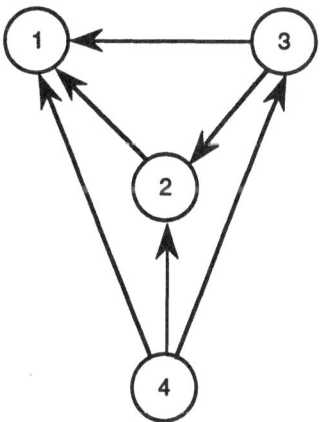

Figure 7.3

Conclusion: Binary relations can be represented by diagrams consisting of nodes and directed edges called digraphs. The individual nodes represent elements of the set on which the relation is defined and the individual edges joining the nodes each represent an element of the relation.

Worked example 7.3

Construct the digraphs of each of the following relations:
(a) R = {(0, 0), (1, 1), (2, 2), (1, 0), (2, 1), (0, 2), (0, 3)}
(b) R = {(0, 0), (1, 2), (2, 3), (3, 4), (1, 1), (2, 2), (3, 3)}
(c) R = {(a, a), (a, b), (a, c), (b, a), (b, c), (d, d)}
(d) R = {(0, 0), (1, 1), (2, 2), (2, 3), (3, 3), (3, 2)}

Solution:
(a) See Figure 7.4.
(b) See Figure 7.5.
(c) See Figure 7.6.
(d) See Figure 7.7.

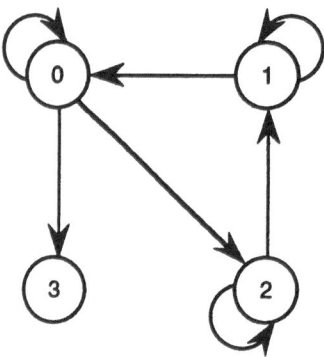

Figure 7.4

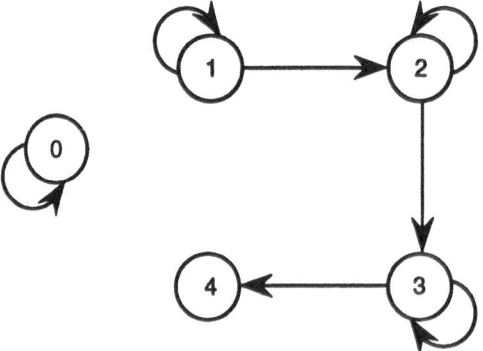

Figure 7.5

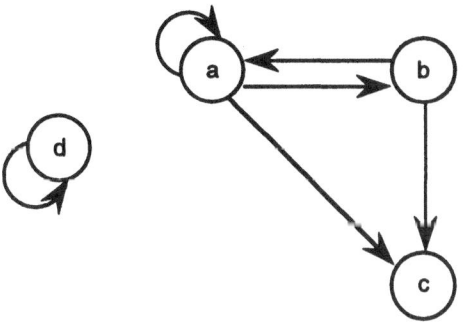

Figure 7.6

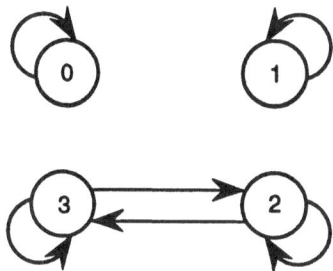

Figure 7.7

Worked example 7.4

Find the explicit relation represented by each of the following digraphs:
(a) Figure 7.8;
(b) Figure 7.9;
(c) Figure 7.10;
(d) Figure 7.11.

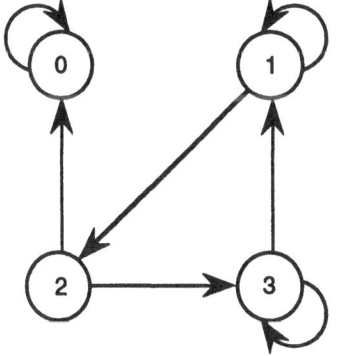

Figure 7.8

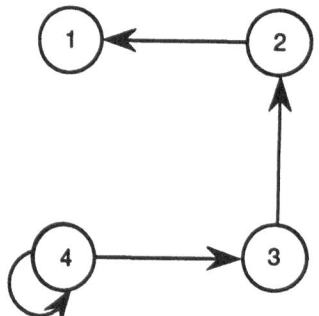

Figure 7.9

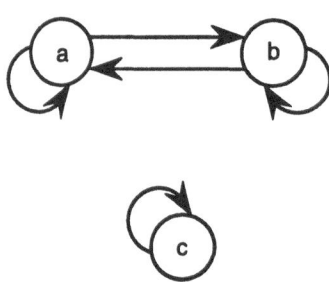

Figure 7.10

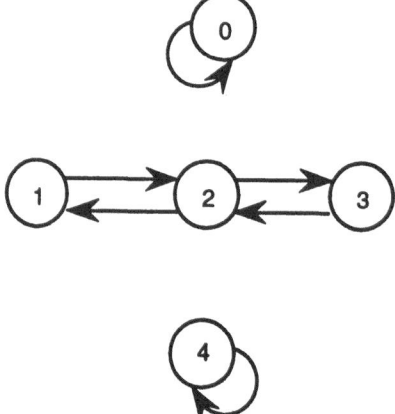

Figure 7.11

Solution:
(a) R = {(0, 0), (1, 1), (1, 2), (2, 0), (2, 3), (3, 1), (3, 3)}
(b) R = {(2, 1), (3, 2), (4, 3), (4, 4)}

(c) R = {(a, b), (b, a), (a, a), (b, b), (c, c)}
(d) R = {(0, 0), (1, 2), (2, 1), (2, 3), (3, 2), (4, 4)}

Exercises

7.3 Construct the digraphs of each of the following relations:
(a) R = {(A, A), (C, C), (D, D), (C, A), (D, C), (A, D), (A, B)}
(b) R = {(00, 01), (00, 10), (00, 11), (01, 10), (01, 11), (10, 11)}
(c) R = {(x, x), (x, y), (y, x), (y, y), (z, z)}
(d) R = {(0, 0), (1, 1), (2, 2), (3, 3), (4, 4)}

7.4 Find the explicit relation represented by each of the following digraphs:
(a) Figure 7.12;
(b) Figure 7.13;
(c) Figure 7.14;
(d) Figure 7.15.

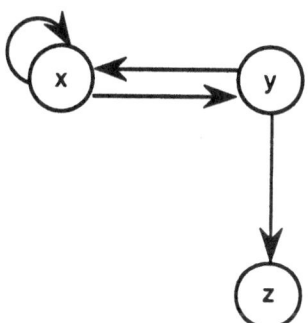

Figure 7.12

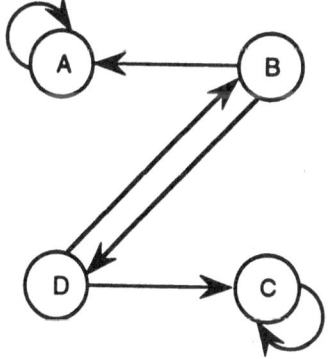

Figure 7.13

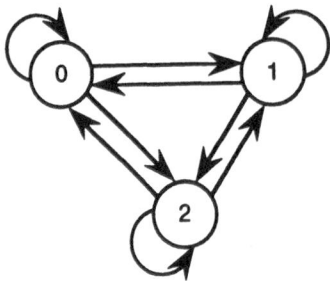

Figure 7.14

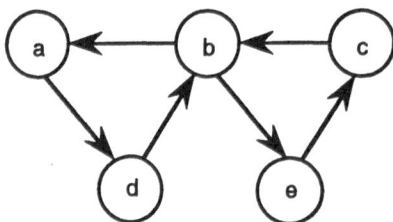

Figure 7.15

Order relations

Order

The two words **data** and **information** are often used interchangeably to the confusion of both because the concepts of data and information, though interdependent, are nonetheless distinct. Data and information can be likened to a received stimulus and the knowledge of its reception respectively. Prick your finger with a pin – that is the data. **Ouch!** that is the information. Information is data within a given context. For example, the string of characters

M123PHD

is a datum until we are told that it is to be found on the number plate of a car. As soon as this context is made evident we realize that the car was first registered in Huddersfield (PHD) some time during the year August 1994 to July 1995 (M). That is information – data within a context and manipulating data within a contextual framework is known as information processing.

Fundamental to information processing is the notion of **order**. The numbers

5, 3, 1, 2, 4

are just numbers – data. The fact that 1 is the smallest number and 5 is the largest number is information, and this information was gained by imposing a context of order on the data.

Numerical order is not the only type of order possible. Letters of the alphabet can be arranged in alphabetical order, words in a dictionary are ordered, the colours of the rainbow are traditionally ordered from red to violet and children can be ordered according to size. In each case there is a condition external to the data that when applied to the data set it puts the data in the desired order.

Order is imposed on a data set by demanding that the data satisfy a relationship that expresses the desired order. If we use variables to represent the various data values then imposing the relationship enables us to construct a relation – an **order relation**. For example, given the set of numbers

$$D = \{4, 6, 2\}$$

we desire to order these numbers in pairs where the first element of each pair is less than or equal to the corresponding second element. Imposing this relationship between elements of D produces the relation R, where

$$R = \{(x, y): x, y \in D \text{ and } x \leq y\}$$

This is given explicitly as

$$R = \{(4, 4), (6, 6), (2, 2), (4, 6), (2, 4), (2, 6)\}$$

with the digraph in Figure 7.16.

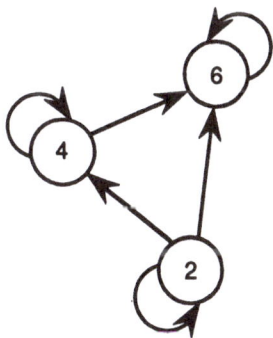

Figure 7.16

From the digraph, drawn purposefully so that all the arrows point in an upwardly direction, we can see that the smallest number, 2, is at the bottom of the digraph, the greatest number, 6, is at the top and the number 4, the middle number, is in the middle of the digraph. The digraph is displaying the order of – the relationship between – the numbers. Notice that all of the elements of D satisfy the relationship with themselves because any number is less than or equal to itself.

This relation is called an order relation because it imposes order on the data set D and permits, among other things, the greatest and the least elements of D to be readily identified. While this may seem to be overstating the problem of finding the greatest and the least of three numbers it is the principle that is important. The principle can be extended to finding the greatest and the least of any number of randomly arranged numbers.

If E is the data set of sets

E = { {apple, pear}, {apple, banana}, {apple, pear, banana},{apple, pear, banana, orange} }

the relation S defined as

S = {(X, Y): X, Y ∈ E and X ⊇ Y}

is another order relation. To simplify the representation we shall denote the elements of E as

A = {apple, pear}
B = {apple, banana}
C = {apple, pear, banana}
D = {apple, pear, banana, orange}

The relation S is then given explicitly as

S = {(A, A), (B, B), (C, C), (D, D), (A, C), (B, C), (A, D), (B, D), (C, D)}

with the digraph in Figure 7.17.

Again, the digraph is drawn purposefully so that all the arrows point in an upwardly direction. Notice that in this digraph A and B are not joined because

(A, B) ∉ S and (B, A) ∉ S

Furthermore, sets A and B are both at the bottom of the digraph, indicating that they are the sets with the smallest cardinality, and D is the topmost set, indicating that it is the set with the highest cardinality. Here ordering the sets has imposed the notion of increasing cardinality.

Total and partial order
The order relation R is called a **total order relation** on set D, whereas the

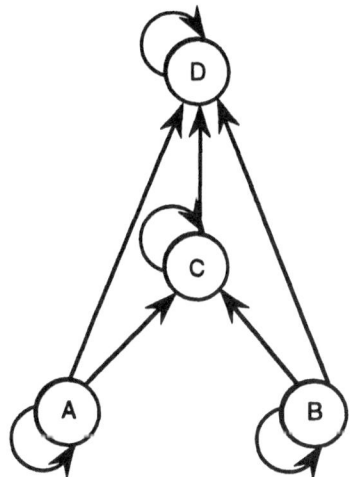

Figure 7.17

relation S is called a **partial order relation**. on set E. In a total order relation every element of the data set is uniquely associated with every other
element of the data set. Diagrammatically this means that every node of the
digraph is connected by a single directed edge to every other node. In a partial order relation this is not so. As we see from the digraph of relation S,
node A is not connected to node B because A and B do not mutually satisfy
the relationship that is used to create the relation.

*Conclusion: An order relation defined on a set D has a digraph which dis-
plays the order imposed on the elements of the set. The order displayed is
either total or partial.*

Worked example 7.5

Describe each of the following order relations explicitly:
(a) $R = \{(x, y): x, y \in \{1/\sqrt{2}, \sqrt{3}/2, \sqrt{(2/3)}, 1/\sqrt{3}\}$ and $x^2 \le y^2\}$
(b) $R = \{(a, b): a, b \in \{$cart, car, bicycle, train, aeroplane$\}$ and a was
     invented at the same time or before $b\}$
(c) $R = \{(p, q): p, q \in \{$Aristotle, Boole, Copernicus, Darwin$\}$ and p is a
     contemporary of or predates $q\}$
(d) $R = \{(s, t): s, t \in \{$brackets, multiplication, addition$\}$ and s either oper-
     ates at the same time or takes precedence over $t\}$

Solution:
(a) $\{(1/\sqrt{3}, 1/\sqrt{3}), (1/\sqrt{3}, 1/\sqrt{2}), (1/\sqrt{3}, \sqrt{(2/3)}), (1/\sqrt{3}, \sqrt{3}/2), (1/\sqrt{2}, 1/\sqrt{2}),$
     $(1/\sqrt{2}, \sqrt{(2/3)}), (1/\sqrt{2}, \sqrt{3}/2), (\sqrt{(2/3)}, \sqrt{(2/3)}), (\sqrt{(2/3)}, \sqrt{3}/2), (\sqrt{3}/2,$
     $\sqrt{3}/2)\}$

(b) {(cart, cart), (cart, train), (cart, bicycle), (cart, car), (cart, aeroplane), (train, train), (train, bicycle), (train, car), (train, aeroplane), (bicycle, bicycle), (bicycle, car), (bicycle, aeroplane), (car, car), (car, aeroplane), (aeroplane, aeroplane)}

(c) {(Aristotle, Aristotle), (Aristotle, Copernicus), (Aristotle, Darwin), (Aristotle, Boole), (Copernicus, Copernicus), (Copernicus, Darwin), (Copernicus, Boole), (Darwin, Darwin), (Darwin, Boole), (Boole, Boole)}

(d) {(brackets, brackets), (brackets, multiplication), (brackets, addition), (multiplication, multiplication), (multiplication, addition), (addition, addition)}

Worked example 7.6

For each of the following order relations, find an assertion that enables the construction of the relation and write down the formal description of the relation:

(a) $R = \{(0, 0), (0, 1), (0, 2), (0, 3), (0, 4), (1, 1), (1, 2), (1, 3), (1, 4), (2, 2), (2, 3), (2, 4), (3, 3), (3, 4), (4, 4)\}$

(b) $R = \{(x, x), (x, w), (x, b), (x, a), (w, w), (w, b), (w, a), (b, b), (b, a), (a, a)\}$

(c) $R = \{(dgt, dgt), (dgt, hu), (dgt, k), (hu, hu), (hu, k), (k, k)\}$

(d) $R = \{(mercury, mercury), (mercury, earth), (mercury, mars), (mercury, venus), (earth, earth), (earth, mars), (earth, venus), (mars, mars), (mars, venus), (venus, venus)\}$

Solution:

(a) $R = \{(n, m): n, m \in \{0, 1, 2, 3, 4\}$ and $n \leq m\}$

(b) $R = \{(s, t): s, t \in \{a, b, w, x\}$ and s does not come before t in the alphabet$\}$

(c) $R = \{(<p>, <q>): <p>, <q> \in \{dgt, hu, k\}$ and len$<p> \geq$ len$<q>\}$

(d) $R = \{(x, y): x, y \in \{mercury, earth, mars, venus\}$ and x is as close as or closer to the sun as $y\}$

Exercises

7.5 Describe each of the following order relations explicitly:

(a) $R = \{(x, y): x, y \in \{-3, -2, -1, 0, 1, 2, 3\}$ and $x - y \geq 0\}$

(b) $R = \{(a, b): a, b \in \{2, 4, 6, 8, 10, 12\}$ and $a/b = n$ where $n \in \mathcal{N}\}$

(c) $R = \{(p, q): p, q \in \{alpha, beta, gamma, delta\}$ and p is not before q in the Greek alphabet$\}$

(d) $R = \{(s, t): s, t \in \{10\ mm, 1\ cm, 2\ m, 0.1\ km\}$ and s is not a larger distance than $t\}$

7.6 For each of the following order relations, find an assertion that enables the construction of the relation and write down the formal description of the relation:

(a) R = {(bat, cat), (bat, fat), (bat, hat), (cat, fat), (cat, hat), (fat, hat), (bat, bat), (cat, cat), (fat, fat), (hat, hat)}

(b) R = {(2, 2), (2, 4), (2, 8), (2, 12), (4, 4), (4, 8), (4, 12), (8, 8), (12, 12)}

(c) R = {({a}, {a}), ({b}, {b}), ({a}, {a, b}), ({b}, {a, b}), ({a, b}, {a, b}), (∅, ∅), (∅, {a}), (∅, {b}), (∅, {a, b})}

(d) R = {(00, 00), (00, 01), (00, 10), (00, 11), (01, 01), (01, 10), (01, 11), (10, 10), (10, 11), (11, 11)}

Relations without express relationships

We have defined a binary relation as a set of ordered pairs in which the elements of each ordered pair satisfy some expressed relationship. Sometimes we are faced with a set of ordered pairs without an explicit relationship being given. Clearly the set of ordered pairs can be regarded as a relation, but can we determine whether or not it is an order relation? Every binary relation can be represented by a digraph but an order relation has a digraph of a particular type. For example, the relation V:

$$V = \{(B, B), (C, C), (D, D), (A, C), (C, A), (B, C), (A, D), (C, D)\}$$

has the digraph shown in Figure 7.18.

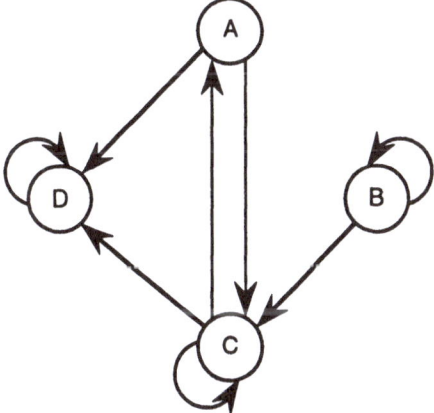

Figure 7.18

However, V is not an order relation for three reasons. First, the element

(A, A)

is absent and in every order relation defined on a set every element of the set must satisfy the relationship with itself. Secondly, the elements

(A, C) and (C, A)

are both present and in an order relation any pair of elements of the set on which the relation is defined can only be related one way. That is the presence of (A, C) indicates that A is ordered relative to C in which case, in an order relation, C cannot also be ordered relative to A. Finally, the element

(B, D)

is absent for a reason that is more subtle than the other two and which will be considered below.

From this discussion it is clear that we need to know more about the properties of relations in general before we can determine whether or not a given relation is an order relation.

Reflexivity, transitivity and antisymmetry
Let R be a relation, defined on data set D where

$$R = \{(x, y): x, y \in D \text{ and } P(x, y)\}$$

and where the open sentence $P(x, y)$ represents the relationship that must be satisfied by x and y for their inclusion in the relation.

Reflexivity
If, $\forall x \in D$, $(x, x) \in R$ then R is called a **reflexive** relation. In a reflexive relation defined on data set D, every element of D is related to itself and the digraph of a reflexive relation is distinguished by the fact that every node has a loop because every node is related to itself (Figure 7.19).

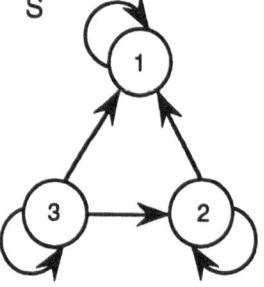
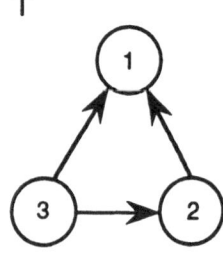

Figure 7.19

For example, the relation

$$S = \{(a, b): a, b \in \{1, 2, 3\} \text{ and } a \geq b\}$$

is reflexive, whereas the relation

$$T = \{(a, b): a, b \in \{1, 2, 3\} \text{ and } a > b\}$$

is not reflexive.

Antisymmetry

If, $\forall x$ and $y \in D$, $(x, y) \in R \rightarrow (y, x) \notin R$ then R is called an **antisymmetric** relation. The digraph of an antisymmetric relation is distinguished by pairs of connected nodes being joined together by only one line (Figure 7.20).

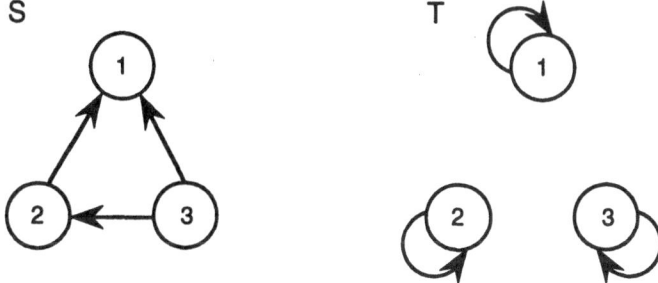

Figure 7.20

For example, the relation

$$S = \{(a, b): a, b \in \{1, 2, 3\} \text{ and } a > b\}$$

is antisymmetric, whereas the relation

$$T = \{(a, b): a, b \in \{1, 2, 3\} \text{ and } a = b\}$$

is not antisymmetric.

Transitivity

If, $\forall x, y$ and $z \in D$, $(x, y) \in R \land (y, z) \in R \rightarrow (x, z) \in R$ then R is called a **transitive** relation. The digraph of a transitive relation is distinguished by triangular connections that demonstrate how any set of three elements of the data set are interrelated (Figure 7.21).

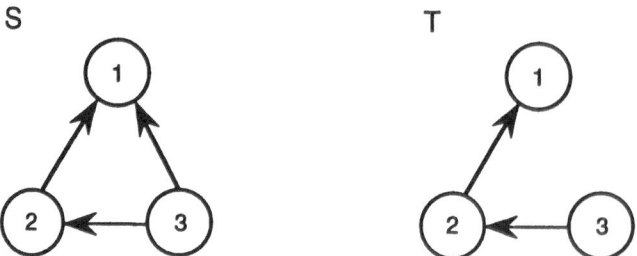

Figure 7.21

For example, the relation

$$S = \{(a, b): a, b \in \{1, 2, 3\} \text{ and } a > b\}$$

is transitive, whereas the relation

$$T = \{(a, b): a, b \in \{1, 2, 3\} \text{ and } a = b + 1\}$$

is not transitive.

Order relation

An order relation is any relation that is reflexive, transitive and antisymmetric, and it is precisely these properties of an order relation that permits us to conclude that the relation T is not an order relation (Figure 7.22).

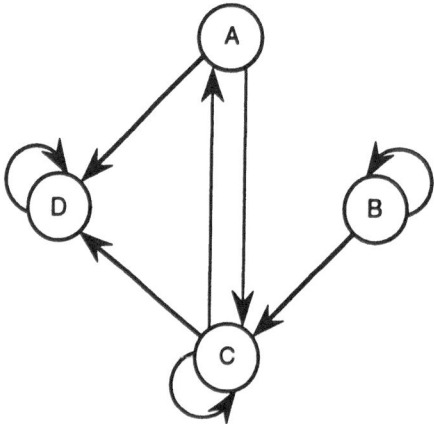

Figure 7.22

The absence of

(A, A)

shows that V is not a reflexive relation. The presence of

(A, C) and (C, A)

shows that T is not an antisymmetric relation and the absence of

(B, D)

shows that T is not a transitive relation. Hence, T fails to be an order rela
tion on all three counts.

*Conclusion: An order relation is a reflexive, transitive and antisymmetric
binary relation.*

Worked example 7.7

Which of the following relations are reflexive?
(a) $R = \{(1, 1), (-1, -1), (1, 2), (-1, 2), (2, 1), (2, -1)\}$
(b) $R = \{(x, y): x, y \in \mathcal{N} \text{ and } x \leq y\}$
(c) $R = \{(a, a), (a, aa), (a, aaa), (aa, aa), (aa, aaa), (aaa, aaa)\}$
(d) $R = \{(x, y): x, y \in \mathcal{N} \text{ and } x/y \notin \mathcal{N}\}$

Solution:
(a) $R = \{(1, 1), (-1, -1), (1, 2), (-1, 2), (2, 1), (2, -1)\}$ is not reflexive
because (2, 2) is absent.
(b) $R = \{(x, y): x, y \in \mathcal{N} \text{ and } x \leq y\}$ is reflexive because $x \leq x$ for all x.
(c) $R = \{(a, a), (a, aa), (a, aaa), (aa, aa), (aa, aaa), (aaa, aaa)\}$ is reflexive.
(d) $R = \{(x, y): x, y \in \mathcal{N} \text{ and } x/y \notin \mathcal{N}\}$ is not reflexive because $x/x \in \mathcal{N}$
for all x.

Worked example 7.8

Which of the following relations are transitive?
(a) $R = \{(p, q), (p, r), (p, s), (q, r), (q, s)\}$
(b) $R = \{(x, y): x, y \in \{\text{people}\} \text{ and } x \text{ is older than } y\}$
(c) $R = \{(00, 01), (00, 10), (00, 11), (01, 01), (01, 10), (01, 11), (10, 10),$
$(10, 11)\}$
(d) $R = \{(p, q): p, q \in \{\text{books}\} \text{ and } p \text{ has more pages than } q\}$

Solution:
(a) R = {(p, q), (p, r), (p, s), (q, r), (q, s)} is transitive despite the absence of (s, r) or (r, s).
(b) R = {(x, y): x, y ∈ {people} and x is older than y} is transitive because if a is older than b and b is older than c then a is older than c.
(c) R = {(00, 01), (00, 10), (00, 11), (01, 01), (01, 10), (01, 11), (10, 10), (10, 11)} is transitive for reasons similar to (a).
(d) R = {(p, q): p, q ∈ {books} and p has more pages than q} is transitive for reasons similar to (b).

Worked example 7.9

Which of the following relations are antisymmetric?
(a) R = {(a, b), (a, c), (a, d), (b, c), (b, d), (b, a)}
(b) R = {(x, y): x, y ∈ {w: w is a word in a dictionary} and x contains as many letters as y}
(c) R = {(1, 2), (2, 1), (1, 3), (3, 1), (1, 4), (4, 1), (1, 5), (5, 1)}
(d) R = {(p, q): p, q ∈ $\mathcal{N}$ and p/q is an equivalent fraction to 2/3}

Solution:
(a) R = {(a, b), (a, c), (a, d), (b, c), (b, d), (b, a)} is not antisymmetric because of the existence of (a, b) and (b, a).
(b) R = {(x, y): x, y ∈ {w: w is a word in a dictionary} and x contains as many letters as y} is not antisymmetric because if a and b satisfy the assertion so do b and a.
(c) R = {(1, 2), (2, 1), (1, 3), (3, 1), (1, 4), (4, 1), (1, 5), (5, 1)} is not antisymmetric because, for example, both (1, 2) and (2, 1) are elements of the relation.
(d) R = {(p, q): p, q ∈ $\mathcal{N}$ and p/q is an equivalent fraction to 2/3} is antisymmetric because if a/b is equivalent to 2/3 then b/a is not.

Worked example 7.10

Which of the following relations are order relations?
(a) R = {(1, 1), (1, 2), (1, 3), (1, 4), (2, 2), (2, 3), (2, 4), (3, 3), (3, 4), (4, 4)}
(b) R = {(w, w), (x, x), (a, a), (b, b), (w, b), (x, a) (x, b), (b, a)}
(c) R = {(spring, summer), (spring, autumn), (spring, winter), (summer, autumn), (summer, winter), (autumn, winter), (spring, spring), (summer, summer), (autumn, autumn)}
(d) R = {(A, B): A, B ∈ {first son, second son, father, mother} and A is not older than B}

Solution:

(a) R = {(1, 1), (1, 2), (1, 3), (1, 4), (2, 2), (2, 3), (2, 4), (3, 3), (3, 4), (4, 4)} is an order relation because it is reflexive, transitive and antisymmetric.

(b) R = {(w, w), (x, x), (a, a), (b, b), (w, b), (x, a) (x, b), (b, a)} is not an order relation because, owing to the absence of (w, a), it is not transitive.

(c) R = {(spring, summer), (spring, autumn), (spring, winter), (summer, autumn), (summer, winter), (autumn, winter), (spring, spring), (summer, summer), (autumn, autumn)} is not an order relation because, owing to the absence of (winter, winter), it is not reflexive.

(d) R = {(A, B): A, B ∈ {first son, second son, father, mother} and A is not older than B} is an order relation despite not knowing the relative ages of the father and the mother.

Exercises

7.7 Which of the following relations are reflexive?
   (a) R = {(up, up), (up, down), (left, left), (left, right)}
   (b) R = {(X, Y): A ⊇ X, Y and X ⊇ Y}
   (c) R = {(1, 1), (2, 2), (1, 2), (1, 3), (3, 3)}
   (d) R = {(x, y): $x, y \in \mathcal{N}$ and $x/y \neq 1$}}

7.8 Which of the following relations are transitive?
   (a) R = {(a, b), (a, c), (a, d), (b, c), (b, d)}
   (b) R = {(x, y): $x, y \in$ {people} and x is related by marriage to y}
   (c) R = {(000, 001), (000, 010), (000, 011), (001, 010), (001, 011), (010, 010), (010, 011)}
   (d) R = {(p, q): $p, q \in$ {days of the week} and p comes before q}

7.9 Which of the following relations are antisymmetric?
   (a) R = {(1, 2), (1, 3), (1, 4), (2, 3), (2, 4), (2, 1)}
   (b) R = {(x, y): $x, y \in$ {w: w is a word in a dictionary} and x is listed before y}
   (c) R = {(l, m), (m, l), (l, n), (n, l), (l, p), (p, l), (l, q), (q, l)}
   (d) R = {(p, q): $p, q \subset \mathcal{N}$ and $p = 5/q$}

7.10 Which of the following relations are order relations?
   (a) R = {(bat, cat), (bat, fat), (bat, hat), (cat, fat), (cat, hat), (fat, hat), (bat, bat), (fat, fat), (hat, hat)}
   (b) R = {(2, 2), (2, 4), (2, 8), (2, 12), (4, 4), (4, 8), (4, 12), (8, 8), (12, 12)}
   (c) R = {({a}, {a}), ({b}, {b}), ({a}, {a, b}), ({b}, {a, b}), ({a, b}, {a, b})}
   (d) R = {(00, 00), (00, 01), (00, 11), (01, 01), (01, 10), (01, 11), (10, 10), (10, 11), (11, 11)}

Representative elements of a set

The other day I went into my local corner store and bought a new brand of chocolate bar. When I sank my teeth into it my face expressed disappointment because it contained peanuts and I am not particularly partial to peanuts and chocolate mixed together. I do not need to buy another bar of that particular brand to see whether or not I shall enjoy it because I know that it will taste the same as the one I just bought. The original bar was **representative** of the entire production line, and when you have tasted one you have tasted them all. Indeed, the world's entire stock of chocolate bars can be divided into those that contain a mixture of peanuts and chocolate and those that do not. As far as those that do, any one of them is a representative of a type that I do not enjoy eating.

This idea of the representation of a type is very important in information processing. Consider the data set of currency units

D = {dollar, yen, pound, franc, mark}

Using the assertion

x is a European currency unit where $x \in$ D

it is possible to partition the set D into the two subsets

{dollar, yen} and {pound, franc, mark}

Clearly, the partition, effected by applying the assertion, distinguishes those currency units that are European from those that are not.

If we now consider the property of **belonging to a particular subset** then dollar and yen both possess the same property as they are both members of the same subset: they are both non-European currency units. By the same reasoning we can say that pound, franc and mark also possess the same property as they too all belong to the same subset: they are all European currency units. However, franc and dollar do not possess the same property as they belong to different subsets, the first being a European and the second a non-European currency unit.

The property of **belonging to** can be used to define the relationship

x and y belong to the same subset

and this can be used to form a relation P defined on set D:

P = {(x, y): x , $y \in$ D and x and y belong to the same subset}

which is given explicitly as

P = {(d, d), (y, y), (d, y), (y, d), (p, p), (f, f), (m, m), (p, f), (f, p), (p, m), (m, p), (f, m), (m, f)}

where, for brevity, we have used the first letter of the currency unit as a substitute for the full name. This relation has the digraph shown in Figure 7.23.

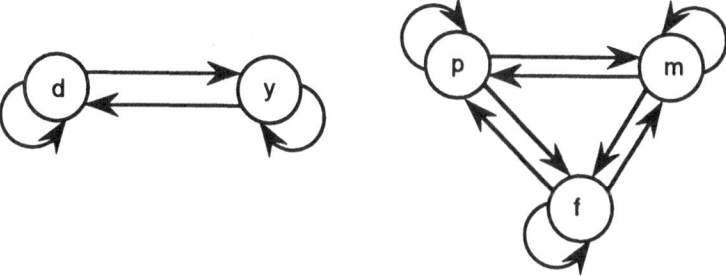

Figure 7.23

Notice that this relation is reflexive. It is also transitive, but it is not anti-symmetric because d and y and p, f and m are joined in pairs by two arrows pointing in opposite directions. Indeed, this relation is not antisymmetric but **symmetric**.

Symmetry
If, $\forall x, y \in D, (x, y) \in R \rightarrow (y, x) \in R$ then R is called a **symmetric** relation. The digraph of a symmetric relation is distinguished by the double connections of any pair of connected nodes (Figure 7.24).

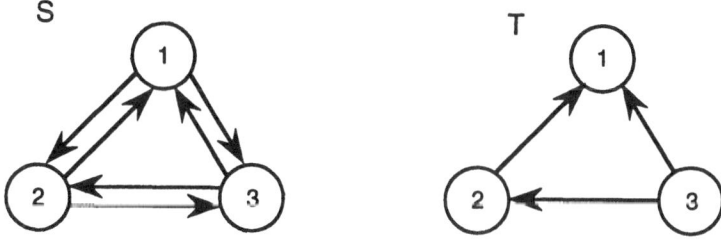

Figure 7.24

For example, the relation

$S = \{(a, b): a, b \in \{1, 2, 3\}$ and $x \neq y\}$

is symmetric, whereas the relation

$$T = \{(a, b): a, b \in \{1, 2, 3\} \text{ and } x > y\}$$

is not symmetric.

Equivalence relations

The relation P formed as a consequence of the partition of set D is reflexive, transitive and symmetric, and any relation that possesses these three properties is called an **equivalence** relation. Indeed, the partition of any set induces an equivalence relation on that set.

The digraphs of equivalence relations are quite distinctive and easy to spot in that they all possess a symmetric structure (Figure 7.25).

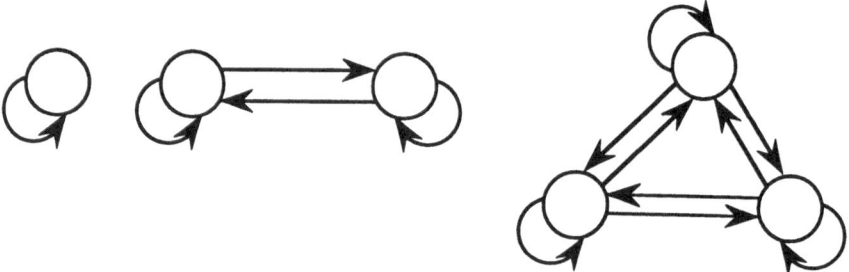

Figure 7.25

Conclusion: An equivalence relation is any relation that is reflexive, symmetric and transitive.

Worked example 7.11

Which of the following relations are symmetric?
(a) $R = \{(a, b): a, b \in \mathcal{N} \text{ and } ab = 5\}$
(b) $R = \{(0, 0), (1, 1), (-1, -1), (2, 2), (-2, -2), (0, 1), (-1, 0), (0, -1),$
     $(2, 0), (0, 2), (-2, 0), (0, -2)\}$
(c) $R = \{(<string1>, <string2>): \text{len}<string1> \neq \text{len}<string2>\}$
(d) $R = \{(a, z), (b, y), (c, x), (d, w), (w, d), (x, c), (y, b), (z, a)\}$

Solution:
(a) $R = \{(a, b): a, b \in \mathcal{N} \text{ and } ab = 5\}$ is symmetric, for example $5 \times 1 = 1 \times 5 = 5$
(b) $R = \{(0, 0), (1, 1), (-1, -1), (2, 2), (-2, -2), (0, 1), (-1, 0), (0, -1),$
     $(2, 0), (0, 2), (-2, 0), (0, -2)\}$ is not symmetric because $(1, 0)$ is not an element of the relation.
(c) $R = \{(<string1>, <string2>): \text{len}<string1> \neq \text{len}<string2>\}$ is symmetric

because if two strings A and B are of unequal lengths then the length of A does not equal the length of B and vice versa.
(d) R = {(a, z), (b, y), (c, x), (d, w), (w, d), (x, c), (y, b), (z, a)} is symmetric.

Worked example 7.12

Which of the following relations are equivalence relations?
(a) R = {(P, Q): P, Q ∈ {all compound propositions} and $P \equiv Q$}
(b) R = {(0, 0), (1, 1), (−1, −1), (0, 1), (1, 0), (−1, 0), (0, −1), (1, −1)}
(c) R = {((a, b), (c, d)): a, b, c, d ∈ N and $ad = bc$}
(d) R − {(a, z), (b, y), (c, x), (x, c), (y, b), (z, a)}

Solution:
(a) R = {(P, Q): P, Q ∈ {all compound propositions} and $P \equiv Q$} is symmetric, transitive and reflexive, therefore it is an equivalence relation.
(b) R = {(0, 0), (1, 1), (−1, −1), (0, 1), (1, 0), (−1, 0), (0, −1), (1, −1)} is not symmetric because of the absence of (−1, 1) therefore it is not an equivalence relation. Also, it is not transitive for the same reason that it is not symmetric.
(c) R = {((a, b), (c, d)): a, b, c, d ∈ $\mathcal{N}$ and $ad = bc$} is an equivalence relation that expresses the equivalence of fractions:

a/b = c/d

(d) R = {(a, z), (b, y), (c, x), (x, c), (y, b), (z, a)} is not reflexive and, therefore, is not an equivalence relation.

Exercises

7.11 Which of the following relations are symmetric:
(a) R = {(a, b): a, b ∈ $\mathcal{N}$ and $a = 3/b$}
(b) R = {(p, p), (q, q), (r, r), (s, s), (t, t), (p, q), (q, p), (r, p), (p, r), (s, p), (p, s), (t, p), (p, t)}
(c) R = {(<string1>, <string2>): len<string1> = len<string2>}
(d) R = {(1, a), (a, 1), (2, b), (b, 3), (c, 3), (3, c), (1, b), (b, 1)}

7.12 Which of the following relations are equivalence relations?
(a) R = {(P, Q): P, Q ∈ {all compound propositions} and the truth table of P is identical to the truth table of Q}
(b) R = {(t, t), (u, u), (v, v), (t, u), (t, v), (u, t), (u, v), (v, t), (v, u)}
(c) R = {(a, b): a, b ∈ $\mathcal{R}$ and $ab = 1$}
(d) R = {(9, 9), (5, 5), (−3, −3), (9, −3), (5, −3), (−3, 9), (−3, 5)}

Equivalence classes

We began by discussing representative elements of a set and, according to the relationship

x and y belong to the same subset of the partition of D

we find that because pound, franc and mark all belong to the same subset then each one of the elements pound, franc and mark can be considered as being representative of the subset to which it belongs. In other words, the elements pound, franc and mark are **equivalent** as representing the particular subset of which they are members, just as dollar and yen are equivalent as representing the particular subset to which they belong. Equivalent representation of a subset is formalized by using the notion of an **equivalence class**.

If an equivalence relation is defined on a set D and a is an element of D then there will a subset of D, each of whose elements are all related to a. This subset is called an equivalence class and is denoted by [a].

Consequently, in our example

[dollar] is the equivalence class {dollar, yen}

because both dollar and yen are related to dollar. Also

[yen] is the same equivalence class {dollar, yen}

because both dollar and yen are also related to yen.

In this respect we write

[dollar] = [yen]

By the same reasoning

[pound] = [franc] = [mark]

because each of these three equivalence classes are the same set

{pound, franc, mark}

Notice that the elements of each equivalence class combine under the relationship of R to form an equivalence relation on each of the subsets of the partition.

Conclusion: An equivalence relation is a reflexive, transitive and symmetric binary relation. Every partition of a set induces an equivalence relation on that set where each subset of the partition is an equivalence class.

Worked example 7.13

Construct the equivalence classes of each of the following relations:
(a) R = {(1, 1), (2, 2), (3, 3), (1, 2), (2, 1)}
(b) R = {(a, a), (b, b), (c, c), (d, d), (e, e), (f, f), (a, b), (b, a), (a, c), (c, a), (b, c), (c, b), (d, e), (e, d)}
(c) R = {(x, y): x, y ∈ {m: m is a member of the University and x is the same age as y}

Solution:
(a) [1] = [2] = {1, 2} and [3] = {3}
(b) [a] = [b] = [c] = {a, b, c}
     [d] = [e] = {d, e} and
     [f] = {f}
(c) [..]   = {...}
     [18] = {all University members who are 18 years old}
     [19] = {all University members who are 19 years old}
     [20] = {all University members who are 20 years old}
     [..]   = {...}

Exercise

7.13 Construct the equivalence classes of each of the following relations:
(a) R = {(w, w), (x, x), (y, y), (z, z), (y, z), (z, y)}
(b) R = {(a, a), (b, b), (c, c), (d, d), (e, e), (f, f), (a, b), (b, a), (a, c), (c, a), (b, c), (c, b), (d, e), (e, d), (d, f), (f, d), (e, f), (f, e)}
(c) R = {(x, y): x, y, ∈ {m: m is a member of the University} and x is in the same class as y}

Trees

Relations can be used to represent the storage of information within a computer. For example, consider the following hierarchical tree that represents a typical directory structure on a working PC (Figure 7.26).

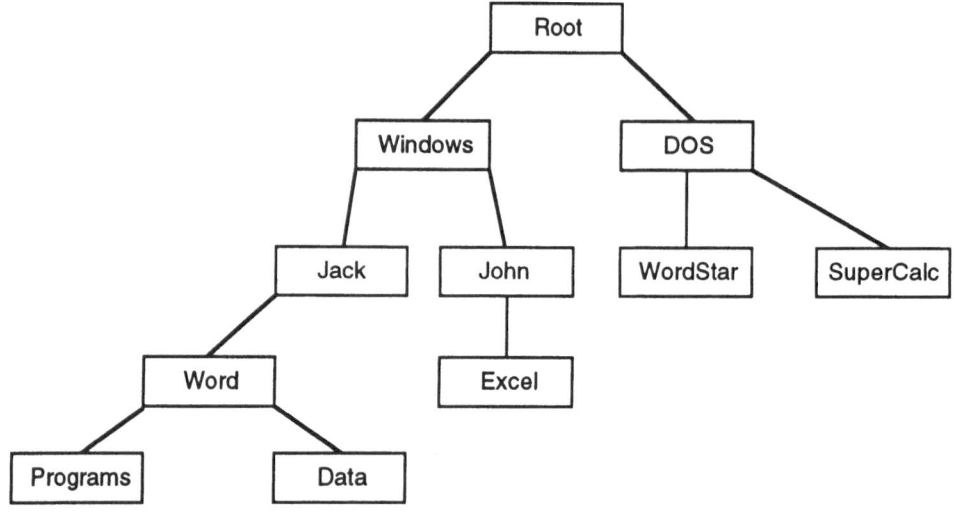

Figure 7.26

This structure can be described using the relationship

    x is a subdirectory of y where $x, y \in D = \{x: x$ is a directory$\}$

where the set of directories is given explicitly as

    D = {Root, DOS, Windows, Wordstar, Supercalc, Jack, John, Word, Excel, Programs, Data}

The structure can also be represented by the relation R, where

    R = {(Root, DOS), (Root, Windows), (DOS, Wordstar), (DOS, Supercalc), (Windows, Jack), (Windows, John), (Jack, Word), (Word, Programs), (Word, Data), (John, Excel)}

Notice that this relation is not reflexive. Indeed, there is not even one node that is related to itself. Such a relation is called **irreflexive**. The relation is, however, antisymmetric but not transitive and this forms the commonality of all trees.

If, given the relation

    $R = \{(x, y): x, y \in D$ and $P(x, y)\}$

R is irreflexive and antisymmetric and further if

    $(a, b) \in R \wedge (b, c) \in R \rightarrow (a, c) \notin R$

we call the relation R a tree.

Every tree has a **root** – a single node from which any other node can be reached by traversing the edges in a given direction indicated by the directed edges. Here the given direction is downwards because the root of the tree is at the top of the diagram. If the root were at the bottom then the given direction would be upwards.

There are many kinds of tree but we shall consider just one, the **binary positional tree**.

Binary positional trees

A binary positional tree is a tree whose nodes have at most two directed edges leading from the node. It is called a positional tree because if two directed edges do lead from a given node then one leads to the left and the other leads to the right. For example, Figure 7.27 shows a binary positional tree, whereas the tree in Figure 7.28 is not a binary positional tree.

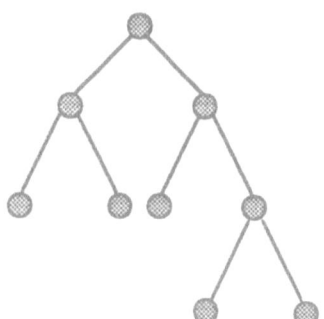

Figure 7.27

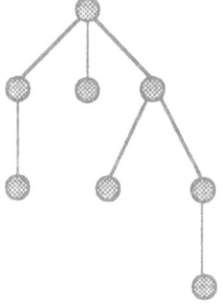

Figure 7.28

Arithmetic expressions

Every arithmetic expression can be represented by a binary positional tree provided a few ground rules are laid out first. We shall represent an expression such as

$$4 + 5$$

by the three-node tree in Figure 7.29.

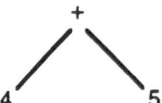

Figure 7.29

Notice that the arithmetic expression is read in the same way that the tree is 'read' – from left to right. This representation can be extended. For example, we represent the expression

$$3 + (4 \times 5)$$

by the tree in Figure 7.30 and the expression

$$(4 + 5) \div (3 - 1)$$

by the tree in Figure 7.31.

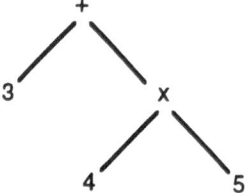

Figure 7.30

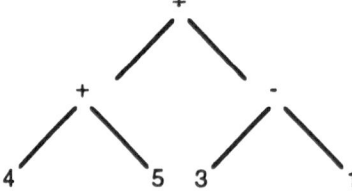

Figure 7.31

These arithmetic expressions exhibit a property common to all arithmetic expressions, namely that there is always a central operator acting between two subexpressions. When using a tree to represent an arithmetic expression then the two subexpressions are to the left and the right of the central operator. Using this principle it is a straightforward matter to construct the tree of quite complicated expressions. For example, to construct the tree of the expression

$$[(4 \times \{5 - 2\}) - (6 \div \{4 - 1\})] + [(8 \div \{6 - 2\}) + (3 \times \{5 + 1\})]$$

notice first of all that the expression is of the form

$$[...] + [...]$$

giving the tree representation in Figure 7.32.

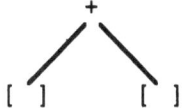

Figure 7.32

The left hand [...] is now seen to be of the form

$$[...] = (...) - (...)$$

Thus the tree is extended to that in Figure 7.33.

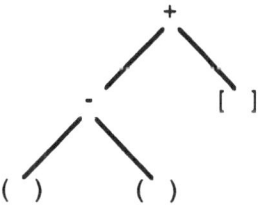

Figure 7.33

Taking each (...) in turn we see that the tree becomes as shown in Figure 7.34.

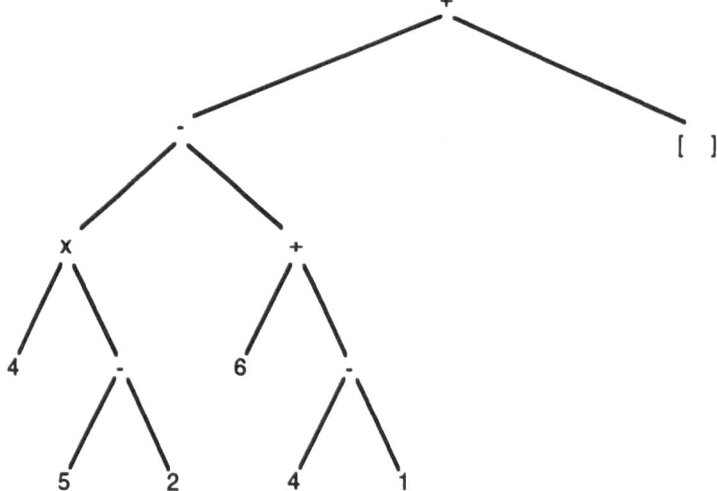

Figure 7.34

Repeating with the right-hand side of the original arithmetic expression we see the completed tree in Figure 7.35.

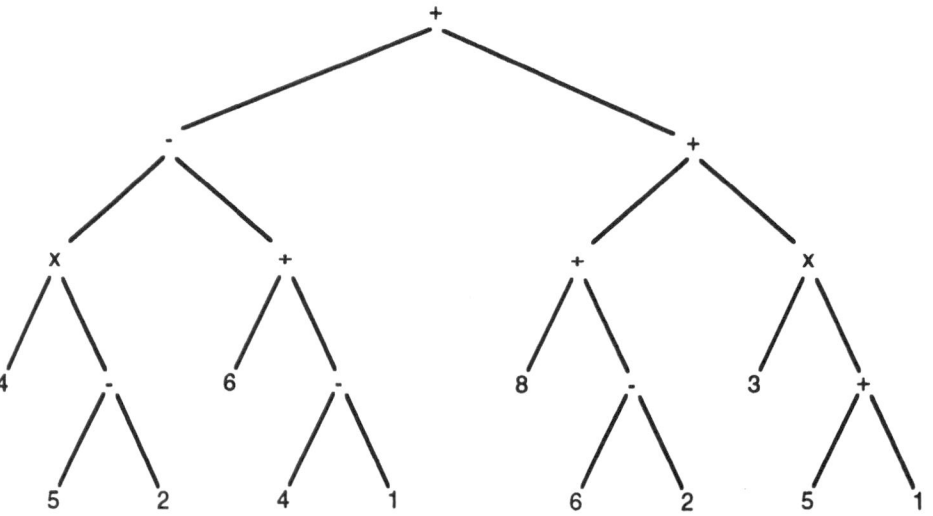

Figure 7.35

Conclusion: A tree is an irreflexive, asymmetric relation with no transitive elements. A binary positional tree is a tree whose nodes can have at most two directed edges leading from them. Such binary positional trees can be used to represent arithmetic expressions.

Worked example 7.14

Construct trees to represent each of the following arithmetic expressions:
(a) $\{(8 - 4) \times 3\} - \{6 \div (2 - 1)\}$
(b) $\{(5 - 2) \times 4\} + \{3 - (2 \div 2)\} - \{5 \times (8 - 5)\}$
(c) $5\{16(5 - 2) - 14/(3 - 1)\}$

Solution:
(a) $\{(8 - 4) \times 3\} - \{6 \div (2 - 1)\}$. Here the central operator is the minus between the two curly brackets. The tree is shown in Figure 7.36.

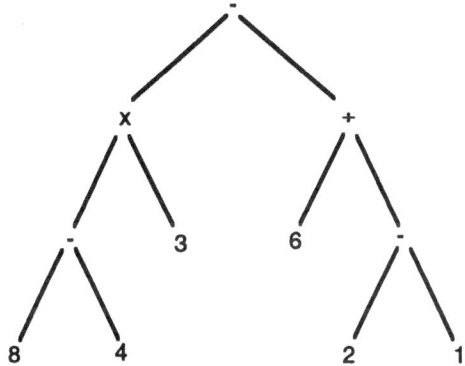

Figure 7.36

(b) $\{(5 - 2) \times 4\} + \{3 - (2 \div 2)\} - \{5 \times (8 - 5)\}$. Here there is a choice of central operator, either the + or the − between the curly brackets. Whichever choice is made will dictate the shape of the tree (Figure 7.37).
(c) $5\{16(5 - 2) - 14/(3 - 1)\}$. Here the arithmetic multiplication is suppressed and the division is changed to the ratio. Convert to the usual form and then construct the tree (Figure 7.38)

$$5\{16(5 - 2) - 14/(3 - 1)\} = 5 \times \{16 \times (5 - 2) - 14 \div (3 - 1)\}$$

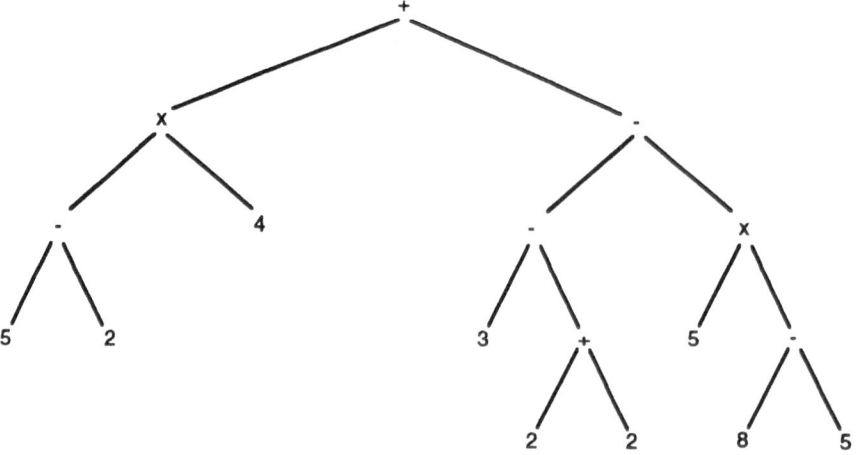

Figure 7.37

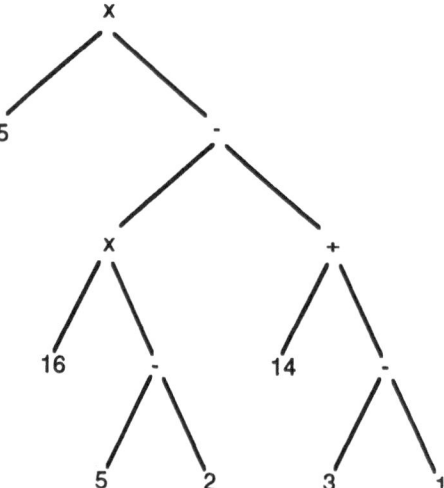

Figure 7.38

Worked example 7.15

Write down the values of the arithmetic expressions represented by each of the trees in:

(a) Figure 7.39;
(b) Figure 7.40;
(c) Figure 7.41.

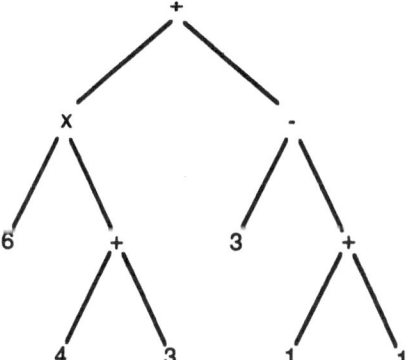

Figure 7.39

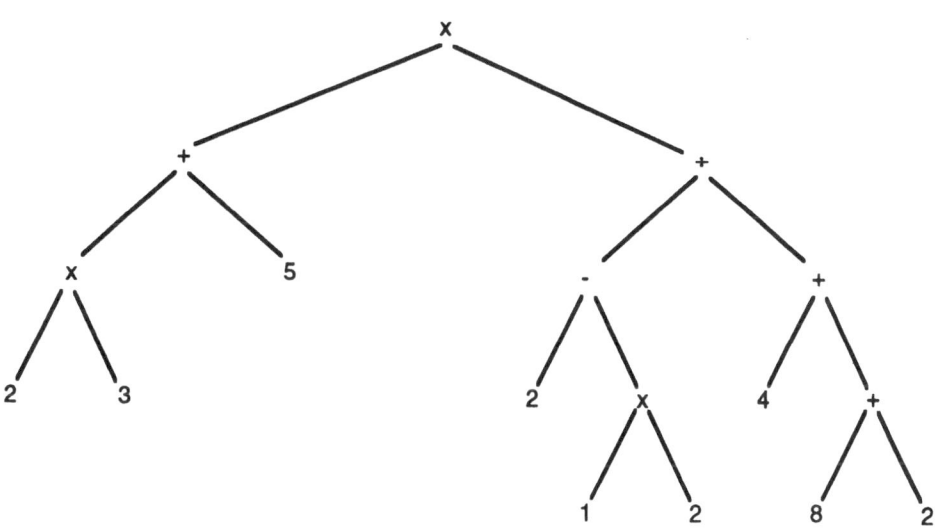

Figure 7.40

Solution:

(a) $(6 \times (4 + 3)) \div (3 - (1 + 1)) = 42$

(b) $((2 \times 3) + 5) \times ((2 - (1 \times 2)) \div (4 + (8 + 2))) = 0$

(c) $2 \times ((9 \times (1 + 2)) + (10 \div (3 + 2))) = 58$

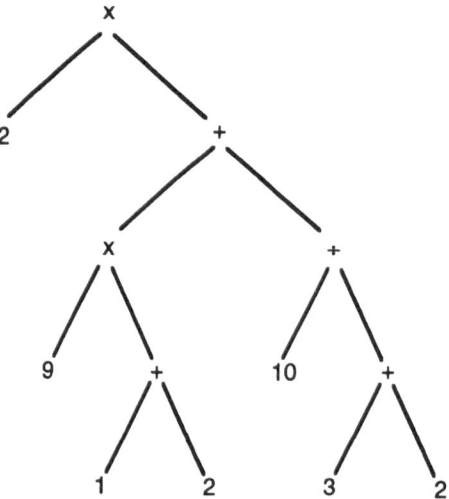

Figure 7.41

Exercises

7.14 Construct trees to represent each of the following arithmetic expressions:

(a) $\{6 \times (3 + 4)\} - \{5 \div (9 - 4)\}$

(b) $\{(3 - 1) \times 2\} - \{8 + (6 \times 3)\} + \{5 \div (10 - 6)\}$

(c) $8\{4/(3 - 1) + 7(9 - 4)\}$

7.15 Write down the values of the arithmetic expressions represented by each of the trees in:

(a) Figure 7.42;

(b) Figure 7.43;

(c) Figure 7.44.

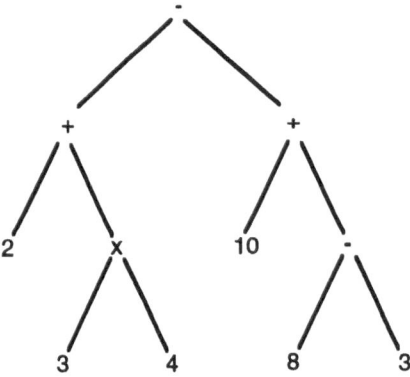

Figure 7.42

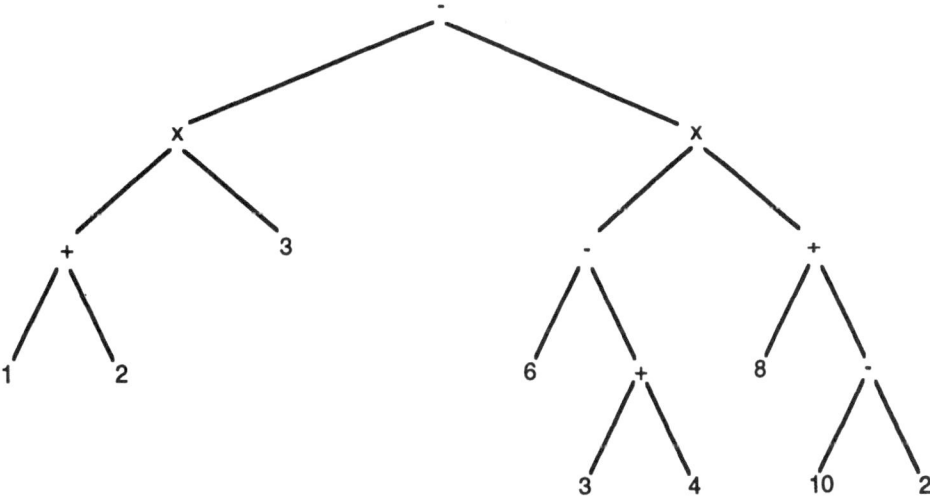

Figure 7.43

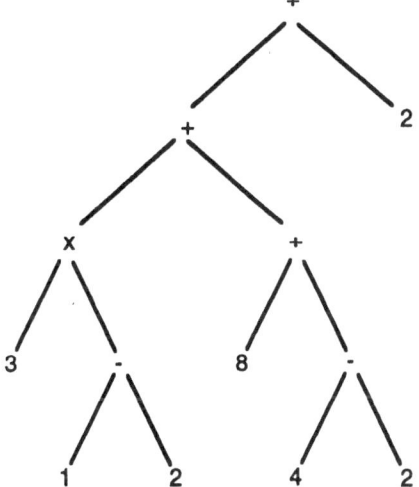

Figure 7.44

Tree searching

We have seen how to create trees as visual representations of arithmetic expressions and, by the same reasoning, we can reproduce an arithmetic expression, and hence its value, from the tree that represents it. However, reproducing the arithmetic expression from the tree that it represents necessitates looking at the tree as a whole – by taking a bird's eye view. Only by doing this are we able to appreciate the structure of the expression and, thereby, replace the brackets which were omitted when the tree was constructed.

Computers are not capable of viewing structures from such a vantage point; they can only consider a structure component by component. Therefore, we need to devise a method whereby the value of the arithmetic expression can be deduced from the tree by traversing the tree from node to node; such is the *modus vivendi* of **tree searching**.

Every tree contains its information at the nodes, and to glean the information stored by traversing the tree it is necessary to visit each node in turn and abstract the information that it stores. This is done by ensuring that each and every node is visited just once. There are just three systematic ways of doing this, and all three start their search at the root of the tree.

The principle involved in searching any tree is to use a recursive process in which every node is considered to be the root of a subtree.

Preorder searching

In a preorder search a node is visited first and the information it contains abstracted. Following this the left subtree is searched and then the right subtree is searched. This recursive procedure can be exemplified by with the following description:

```
BEGIN Search of subtree
        VISIT the root      abstract the information from the root
        Search left subtree
        Search right subtree
END Search of subtree
```

For example, to search the tree that represents the arithmetic expression

$(3 + 4) \times 2$

which has the value 14 (Figure 7.45)

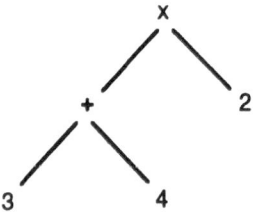

Figure 7.45

we adopt the following procedure:

```
BEGIN Search of tree
        VISIT the root                                            ×
        Search left of ×
        BEGIN Search of subtree
           VISIT the root                 +
           Search left of +
           BEGIN Search of subtree
              VISIT the root              3
              Search left of 3            [nothing left of 3 to search]
              Search right of 3           [nothing right of 3 to search]
           END Search of subtree
           Search right of +
           BEGIN Search of subtree
              VISIT the root              4
              Search left of 4            [nothing left of 4 to search]
              Search right of 4           [nothing right of 4 to search]
           END Search of subtree
        Search right of ×
        BEGIN Search of subtree
           VISIT the root                 2
           Search left of 2               [nothing left of 2 to search]
           Search right of 2              [nothing right of 2 to search]
        END Search of subtree
END Search of tree
```

The sequence of visits results in the collected output from the tree of

   ×+342

The next problem is to make sense of this output, and to do this we make use of what is called **Polish notation** or **prefix notation**, in which the arithmetic operator precedes the two numbers on which it acts.

Polish notation

If p is an arithmetic operator and a and b are two real numbers the arithmetic expression

   a p b

is written in Polish notation as

   pab

Using this notation we can now evaluate the output from the tree as follows (we shall insert brackets to make the evaluation clear):

$$x+342 = x(+34)2$$
$$= x(3 + 4)2$$
$$= x72$$
$$= 7 \times 2$$
$$= 14$$

Postorder searching

A second tree search procedure is the postorder, which is a recursive procedure exemplified by the following description:

```
BEGIN Search of subtree
      Search left subtree
      Search right subtree
      VISIT the root          abstract the information from the root
END Search of subtree
```

Here we see that the visit to a node is made **after** searching both the left and right subtrees to which is the root. A postorder search on the tree yields

```
BEGIN Search of tree
      Search left of X
      BEGIN Search of subtree
         Search left of +
         BEGIN Search of subtree
            Search left of 3      [nothing left of 3 to search]
            Search right of 3     [nothing right of 3 to search]
            VISIT the root     3
         END Search of subtree
         Search right of +
         BEGIN Search of subtree
            Search left of 4      [nothing left of 4 to search]
            Search right of 4     [nothing right of 4 to search]
            VISIT the root     4
         END Search of subtree
         VISIT the root     +
      END Search of subtree
      Search right of X
      BEGIN Search of subtree
         Search left of 2      [nothing left of 2 to search]
         Search right of 2     [nothing right of 2 to search]
         VISIT the root     2
      END Search of subtree
      VISIT the root         X
END Search of tree
```

The sequence of visits results in the collected output from the tree of

$34 + 2\times$

To make sense of this output and to do this we make use of what is called **reverse Polish notation** or **postfix notation**, in which the arithmetic operator follows the two numbers on which it acts.

Reverse Polish notation
If p is an arithmetic operator and a and b are two real numbers the arithmetic expression

a p b

is written in reverse Polish notation as

abp

Using this notation we can now evaluate the output from the tree as follows (we shall insert brackets to make the evaluation clear):

$$
\begin{aligned}
34 + 2\times &= (34+)2\times \\
&= (3 + 4)2\times \\
&= 72\times \\
&= 7 \times 2 \\
&= 14
\end{aligned}
$$

In-order searching
A third tree search procedure is the in-order procedure, which is a recursive procedure exemplified by the following description:

```
BEGIN Search of subtree
      Search left subtree
      VISIT the root        abstract the information from the root
      Search right subtree
END Search of subtree
```

Here we see that the visit to a node is made after searching the left subtree but before searching the right subtrees to which is the root. An in-order search on the tree yields

```
BEGIN Search of tree
      Search left of X
      BEGIN Search of subtree
         Search left of +
```

```
        BEGIN Search of subtree
            Search left of 3          [nothing left of 3 to search]
            VISIT the root            3
            Search right of 3         [nothing right of 3 to search]
        END Search of subtree
        VISIT the root                +
        BEGIN Search of subtree
            Search left of 4          [nothing left of 4 to search]
            VISIT the root            4
            Search right of 4         [nothing right of 4 to search]
        END Search of subtree
        Search right of +
    END Search of subtree
    VISIT the root                    ×
    BEGIN Search of subtree
        Search left of 2              [nothing left of 2 to search]
        VISIT the root                2
        Search right of 2             [nothing right of 2 to search]
    END Search of subtree
    Search right of ×
END Search of tree
```

The sequence of visits results in the collected output from the tree of

$$3 + 4 \times 2$$

This output consists of all the terms in the arithmetic expression in the order in which they appear but without the brackets. If this expression is evaluated according to the precedence rules then it will produce the result

$$\begin{aligned} 3 + 4 \times 2 &= 3 + (4 \times 2) \\ &= 3 + 8 \\ &= 11 \end{aligned}$$

which is the incorrect answer.

We have seen that when a tree is constructed to store an arithmetic expression the structure of the tree reflects the location of brackets within the arithmetic expression. However, the tree itself does not contain any brackets so that when the tree is searched the resulting expression is necessarily bracket-free. An in-order search can result in the value of the expression resulting from the search being different from the one that the expression possessed when the tree was constructed. The only way to guarantee the preservation of the value of an arithmetic expression stored within a tree is to search the tree using either a pre- or a postorder search and to then evaluate the resulting expression using the appropriate Polish or reverse Polish notation.

Worked example 7.16

Perform a pre-, a post- and an in-order search on the trees in each of the following figures and evaluate the results of both the pre- and the postorder searches.

(a)   Figure 7.46;
(b)   Figure 7.47;
(c)   Figure 7.48.

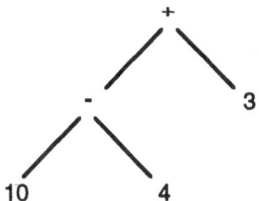

Figure 7.46

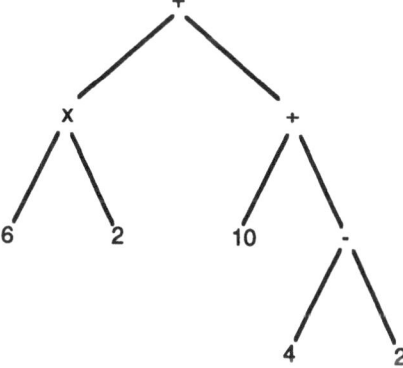

Figure 7.47

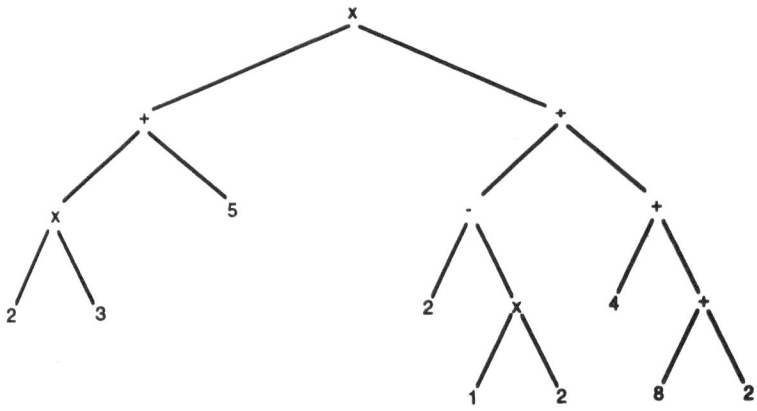

Figure 7.48

Solution:

(a) Preorder

   $\div - 10\ 4\ 3 = \div\ 6\ 3 = 2$

   Postorder

   $10\ 4 - 3 \div = 6\ 3 \div = 2$

   In-order

   $10 - 4 \div 3 = 8^2/_3$

   applying the precedence rules.

(b) Preorder

   $+ \times 6\ 2 \div 10 - 4\ 2 = +\ 12 \div 10\ 2 = +\ 12\ 5 = 17$

   Postorder

   $6\ 2 \times 10\ 4\ 2 - \div + = 12\ 10\ 2 \div + = 12\ 5 + = 17$

   In-order

   $6 \times 2 + 10 \div 4 - 2 = 12 + 2^1/_2 - 2 = 12^1/_2$

   applying the precedence rules.

(c) Preorder

   $\times + \times 2\ 3\ 5 \div - 2 \times 1\ 2 + 4 + 8\ 2\ \ = \times + 6\ 5 \div - 2\ 2 + 4\ 10$

   $= \times\ 11 \div 0\ 14$

   $= \times\ 11\ 0$

   $= 0$

   Postorder

   $2\ 3 \times 5 + 2\ 1\ 2 \times - 4\ 8\ 2 + + \div \times\ = 6\ 5 + 2\ 2 - 4\ 10 + \div \times$

   $= 11\ 0\ 14 \div \times$

   $= 11\ 0 \times$

   $= 0$

In-order

$$2 \times 3 + 5 \times 2 - 1 \times 2 \div 4 + 8 + 2 = 6 + 10 - \frac{1}{2} + 8 + 2$$
$$= 25\frac{1}{2}$$

Exercises

7.16 Perform a pre-, a post- and an in-order search on the trees in each of the following figures and evaluate the results of both the pre- and the postorder searches.
 (a) Figure 7.49;
 (b) Figure 7.50;
 (c) Figure 7.51.

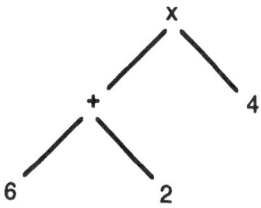

Figure 7.49

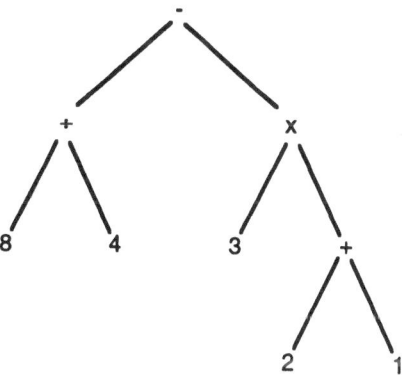

Figure 7.50

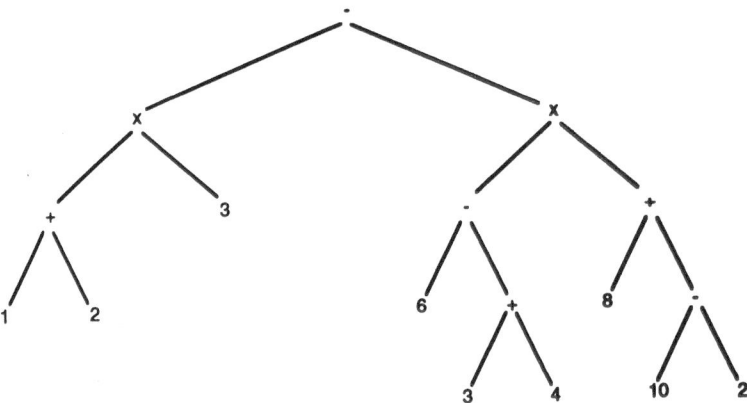

Figure 7.51

Chapter 8

Functions

OBJECTIVES

When you have completed this chapter you will be able to:

□ define a function in terms of a system;

□ define a function as a rule and so determine its domain and range;

□ distinguish between a total and a partial function;

□ generate the graph of a function;

□ appreciate the connection between functions and mappings;

□ distinguish the three types of mapping;

□ find the composition of a collection of functions and decompose a composition;

□ define the inverse of a function and, where appropriate, the inverse function;

□ distinguish between a function and a relation;

□ distinguish between countable and non-countable sequences;

□ manipulate arithmetic and geometric sequences;

□ handle recursive sequences;

□ define functions of two or more variables and know how to curry such functions.

The purpose of a computer system is to produce output. This output may be in the form of a printed document, it may be in the form of a screen display or it may even be in the form of the state of the computer's memory. Whatever form the output may take it is dependent upon, among other things, input. If we are to describe the behaviour of the computer in analytical terms then one of the factors we need to be able to describe is how an output is obtained from an input.

Single variable input systems

Programs and algorithms

Computers do exactly what they have been told to do: no more and no less. Their ability, for example, to display a flashing cursor while they wait for input via a keyboard, a mouse or some other input device is not something innate to the machine itself, it is a consequence of the computer being instructed to do so. This control of the computer by instruction is fundamental to computing and is achieved by issuing a series of coded instructions that the computer is capable of understanding and obeying. The series of coded instructions is called a program and every program starts its life as a desire.

> I wish I could ask my DOS PC to tell me what files I have stored on my hard disk

Fine, we need a program that will do just that. Where do we start? We start by writing down the sequence of operations that must be performed to enable the desire.

1.   Display the C:\> prompt and wait until the instruction DIR followed by <Enter> is issued at the keyboard.
2.   Start the floppy disk drive motor and activate the read head in the disk drive.
3.   Move the read head to that part of the disk that stores the catalogue containing a list of the files on the disk.
4.   Read the catalogue and then convert that information into a screen display.

This list of instructions is called an **algorithm** – the *Concise Oxford Dictionary* defines an algorithm as:

> A process or set of rules used for calculation or problem solving, especially with a computer

Having created the algorithm, albeit a rather crude one, we must then translate the instructions into a code that the computer can understand and act upon. In other words we must program the computer to do this task.

To activate any particular program will require a signal. In our case that signal will be typing in DIR at the C:\> prompt followed by <Enter>. This is called the **input**. The program will then act upon the input, perform the processes that are laid down in the set of coded instructions and finally produce an **output**. In our case the output will be the display of the contents of the disk. We can reduce all this to its essentials by calling it an:

input–process–output system

with the provisos that:

1. an individual input will result in just one output and
2. the same input always produces the same output.

In our case, if we input

```
DIR <Enter>
```

at the C:\> prompt then the computer will produce an output consisting of the directory listing of the hard disk.

```
Volume in drive C is DOS400
Volume Serial Number is 18AF-6931
Directory of C:\
.                 <DIR>           28/02/93   16:51
..                <DIR>           28/02/93   16:51
DOS               <DIR>           28/02/93   16:51
AUTOEXEC    BAT                   28/02/93   17:30
WINDOWS           <DIR>           09/01/94   10:04
```

If we were to make a mistake and input

```
DIT <Enter>
```

at the C:\> prompt then the computer would produce the output consisting of the display:

```
Bad command or file name
```

as it would if we were to type DIS instead of DIT.

Notice that even though we demand a single output for a given input we do not demand that any particular output relates to only one input. Indeed, in our case, while the directory listing relates to the unique input of DIR <Enter>, the output

```
Bad command or file name
```

relates to a plethora of different inputs such as DIT <Enter> or DIS <Enter>.

Every computer program can be described mathematically as a function but first we need to define a system.

Systems
A system consists of

an input

a process that acts on the input
an output that is the result of the processing

The processing unit of the system converts the input into an output. A bicycle is a typical system. The physical structure of the bicycle converts the rotary motion of the pedals into the forward motion of the bicycle. Another system, whose workings are less transparent, is a television remote control device, which processes the push of a button into the display of a television picture. In both cases the intrinsic feature of the processing unit of the system lies in the **effect** of processing and not **how** the processing was effected.

A principal feature of a system is that for a given input it will produce a single output as illustrated by Figure 8.1.

Figure 8.1

This notion of a system can be used to define a mathematical function.

Functions
The processing unit of a system that processes an input into a single output is called a **function**. For example, a familiar function, available on a hand calculator, is the function that produces the reciprocal of the input (Figure 8.2).

Figure 8.2

Another example of a function is a computer program that accepts as input a string of alphanumeric characters and outputs the number of different vowels that the string contains (Figure 8.3).

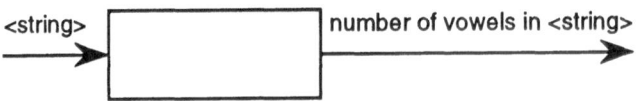

Figure 8.3

Before we can sensibly discuss the working of a particular function it must be given a label or a name. For example, the function that produces the reciprocal of the input could be called **recip** and the function that counts different vowels could be called **vowelcount**. In either case the name is chosen to give some indication of the processing performed by the function. If you have met functions before you may have noticed the overenthusiastic use of the label *f* to denote a function. This is a useful device if brevity is the essence of the notation but it is not a very useful device otherwise.

In addition we need a notation for the input values. For the first example, where the input values are numbers, then a simple letter *x* should suffice. In which case (Figure 8.4) we define the

input as *x*
the process as recip and
the output as recip(x) – to be read as **recip of *x*** or as **the reciprocal of *x*.**

Figure 8.4

In the second example an acceptable notation to denote the input values would be <string>. In this case (Figure 8.5) we define the

input as <string>
the process as vowelcount
the output as vowelcount(<string>)

Figure 8.5

Conclusion: A system consists of an input to a processing unit and an output that is the result of the input being processed where for each input there is a single, consistent output. The processing unit of a system is called a function.

Worked example 8.1

Describe the input, the process and the output for each of the following systems:
(a) counting the number of words in a paragraph;
(b) finding the square of a number;
(c) library enquiry system that gives the location number of a particular title;
(d) listening to the radio.

Solution:
(a) Input the contents of the document paragraph
     Count the number of words
     Output the number of words
(b) Input the number
     Calculate the square of the number
     Output the square of the number
(c) Input the book title
     Find the location number relating to that title
     Output the location number
(d) Input the channel selection
     Convert the radio signal to sound
     Output the programme

Worked example 8.2

Give suitable names to the following functions and suitable labels for their input and output data:
(a) counting the number of words in a paragraph;
(b) finding the square of a number;
(c) library enquiry system that gives the location number of a particular title;
(d) listening to the radio.

Solution:
(a) Input <string>
     Process wordcount
     Output wordcount(<string>)
(b) Input number x
     Process square
     Output square(x)
(c) Input <string>
     Process titlelocation

Output titlelocation(<string>)
(d) Input channel
     Process radio
     Output radio(channel)

Exercises

8.1 Describe the input, the process and the output for each of the following
systems:
(a) measuring the length of a journey;
(b) finding the cube root of a number;
(c) selecting a holiday from a brochure;
(d) spellchecking a document.

8.2 Give suitable names to the following functions and suitable labels for
their input and output data:
(a) measuring the length of a journey;
(b) finding the cube root of a number;
(c) selecting a holiday from a brochure;
(d) spellchecking a document.

Properties of functions

Functions defined as rules
While we are not primarily concerned with how the function converts the
input into an output it will often be the case that for simple functions the
processing can be inferred from the way the output is described in terms of
the input. For example, the function recip has an output defined by

$$\mathrm{recip}(x) = 1/x$$

Clearly this tells us how to derive the output from a given input. We shall
refer to this as a rule which defines the action of the function. Other, more
involved functions can likewise have their actions described by a rule. For
example

vowelcount(<string>) = number of vowels in <string>

Notice, however, that this rule does not indicate how the processing is to be
achieved.

For a complete definition of a function the rule alone is insufficient: we
need supplementary conditions that restrict the inputs to permitted values.
For example, recip accepts real numbers but cannot accept as input the

number 0 because the number 0 does not have a reciprocal – division by zero is not a defined operation. The conditions imposed on the permitted inputs define the **domain** of the function and this is best described using the language of sets.

Domain and range of a function

A function is described as a rule linking an input value to an output value. The function accepts as input an element of set D, called the domain of the function, and processes this input to produce an output that is an element of set R referred to as the **range** of the function. For example, a computer has a program called **name** installed that enables the user to access a database record and print the contents of the first field which contains a person's surname. In the database are four records whose first fields contain the names

Adams, Bassett, Carter and Davis

To activate name the user must input a number from 1 to 4 at the prompt. As an instance, if the user enters the number 3 at the prompt, the program called name causes Carter to be printed on the monitor screen. That is

name(3) = Carter

The program called name is a function whose domain set is

$$D = \{1, 2, 3, 4\}$$

and whose range set is

R = {Adams, Bassett, Carter, Davis}

In conclusion, every function is defined by stating the rule that links the input to the output and by stating the domain and range. The function called name is defined using the following notation:

name: $x \mapsto$ name(x)

where

$$x \in \{1, 2, 3, 4\}$$

and

name(x) $\in$ {Adams, Bassett, Carter, Davis}

The notation

name: $x \mapsto \text{name}(x)$

is referred to as the **signature** of the function.

Notice again that this description of the function called name does not tell us how the function achieves the linking of a domain element with a range element.

Total and partial functions

If the user inputs any keyboard character other than 1, 2, 3 or 4 to the program called name then the program would not be able to produce an output unless an error detection instruction had been instituted that would indicate that an erroneous input had been made.

This problem of catering for erroneous inputs is associated with any computer program and is best dealt with as a separate issue to that of defining the program itself. As far as general functions are concerned it is often more convenient to define the domain of a function to be larger than is necessary and then, separately, to define a restricted subset of the domain as applying to the function.

If the defined domain D of a function contains elements that are not to be used as inputs to the function then the function is called a **partial function** on the domain D. If, however, the subset D′ of D contains just those elements than can be used as inputs to the function then the function is called a **total function** on domain D′. For example, if we define the domain of the program called name to be the set of integers then name is a partial function on the set of integers but a total function on the subset

$$\{1, 2, 3, 4\}$$

Partial functions are met in computing quite frequently. Any function whose input is restricted to a subset of the set of keys on a keyboard is a partial function on the set of all keyboard characters. To convert such a partial function to a total function can be achieved by defining the restricted set as we have just seen or by adding an element to the range called undefined which corresponds to all unallowed input characters. The first of these two options is valid provided it is possible to define those inputs that are allowed and those that are not. The second is equally viable provided we know how to cater for undefined inputs.

Pre- and post-conditions

If a partial function, defined on domain D, is converted to a total function on subset D′ of D by imposing conditions on the elements of D to select those that are valid inputs to the function then these imposed conditions are referred to as the function's **pre-conditions**. For example, consider the function defined as

$$\text{recip: } x \mapsto \text{recip}(x) = 1/x$$

where

$$x \in \mathcal{R}$$

and

$$\text{recip}(x) \in \mathcal{R}$$

Clearly this is a partial function defined on the entire set of real numbers. By restricting the domain of the function to R' where

$$R' = \{x : x \in \mathcal{R} \wedge x \neq 0\}$$

we have imposed a pre-condition that makes recip a total function on R'.

By defining the range of a function, we are imposing conditions on the validity of the output. These conditions are referred to as **post-conditions** and they can be used to ensure that the function has processed the inputs correctly. Indeed, the process of testing a computer program with trial input data is essentially one of testing a function against a set of post-conditions that must be satisfied if the program is processing the trial data correctly.

Ordered pairs and graphs of functions

For every element of the domain the function produces an element of the range. This permits corresponding pairs of domain and range elements to be combined together to form an ordered pair. For example

(2, Bassett)

is one such ordered pair that can be generated from the program called name. The pair is ordered because the domain element appears first and the corresponding range element appears second.

The set G of all possible ordered pairs that can be generated by a function is called the **graph** of the function. For example, the graph of the function name is

An alternative definition of a function is to define it as the graph G. This is a useful device if you wish to consider a function as a special kind of relation. It does mean, however, that if account has to be taken of processing then further definitions are required.

G = {(1, Adams), (2, Bassett), (3, Carter), (4, Davis)}

Notice that the graph of name is a subset of the Cartesian product:

{1, 2, 3, 4} × {Adams, Bassett, Carter, Davis}

The elements x and y of each ordered pair (x, y) of G satisfy the relationship

y is the entry of the Name field of record x

Consequently, not only is G the graph of a function, it is also a relation. We shall discuss the correlation between the graphs of functions and relations later in this chapter.

Mappings

The function called name can be alternatively represented as a process that **links** an element of D with an element of R (Figure 8.6).

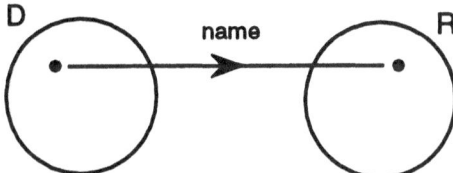

Figure 8.6

Such a linking is called a **mapping**, and the function called name can be said to **map** elements of the domain set to elements of the range set.

Notice that this representation does not give any indication of how the linking is performed – that has to be described elsewhere.

Conclusion: A function can be defined as a rule that links an element of the function's domain to an element of the function's range. If the domain of the function is defined to contain elements that cannot be used as inputs to the function then the function is a partial function on that domain. If the domain is restricted to contain just those elements that are applicable to the function then the function is called a total function on the restricted domain. This restriction can be achieved by using a pre-condition on the function. After a function has been activated a post-condition can be used to check that the function has operated correctly. The graph of the function is then the set of ordered pairs of associated domain and range elements. An alternative description of a function is known as a mapping.

Worked example 8.3

Describe in words how each of the following functions processes the input to produce the output:
(a) counting the number of words in a paragraph;
(b) finding the square of a number;
(c) library enquiry system that gives the location number of a particular title;
(d) listening to the radio.

Solution:
(a) By assuming that every word is separated by a single space then the

number of words can be counted by counting the number of spaces. There will be one more word than spaces so that the number of words will equal the number of spaces plus 1.

(b) The square of a number is found by multiplying the number by itself.

(c) Assuming that all the details referring to a particular book are stored on a single record in a library database the function will search the database to find that record whose title entry matches the input title. The location reference number is then read from the appropriate field in the record.

(d) The radio is tuned to accept a particular channel frequency. This causes the entire radio to vibrate in sympathy with the incoming signal. The vibration is then electronically amplified to the speaker and the result is the sound of the selected programme.

Worked example 8.4

Describe the domain and range of each of the following functions:

(a) counting the number of words in a document

(b) finding the square of a number;

(c) library enquiry system that gives the location number of a particular title;

(d) listening to the radio.

Solution:

(a) Domain $D = \{s: s$ is a string of characters$\}$
    Range $R = \mathcal{N}$

(b) Domain $\mathcal{R}$
    Range $\mathcal{R}$

(c) Domain $D = \{$rec: rec is a record in the library database$\}$
    Range $R = \{$ref: ref is a reference location code$\}$

(d) Domain $D = \{$freq: freq is a radio frequency$\}$
    Range $R = \{p: p$ is a radio programme$\}$

Worked example 8.5

Which of the following are total functions? Give appropriate pre-conditions for those that are partial functions to convert them into total functions:

(a) counting the number of words in a paragraph;

(b) finding the square of a number;

(c) library enquiry system that gives the location number of a particular title;

(d) listening to the radio.

Solution:

(a) This is a partial function. A paragraph is a string of characters that conforms to a strict set of conditions. For example, the string must begin with a capital letter and the string must contain no carriage returns except one at the end. Consequently, if the domain is just the set of strings with no restrictive conditions imposed, the function is a partial function. Imposing these conditions on the original domain will restrict the applicable domain to a subset of the set of strings and convert the function into a total function.

(b) In theory this is a total function because every real number can be multiplied by itself. In practice, however, it is a partial function because it is impossible to compute the square of a number where either the input or the output is an irrational number. To ensure that the function is a total function the domain must be restricted to the rational numbers.

(c) This is a partial function because not all books are going to be in a single library. To convert to a total function the range can be extended to include a *not available* record which caters for the input of a book title that is not in the library.

(d) This is a partial function because there are certain radio frequencies that do not correspond to radio programmes. The partial function can be converted to a total function by restricting the domain to those frequencies that have been allocated to transmitted programmes. Alternatively, we can create a total function by extending the range to include the null program that corresponds to no programme allocated to a particular frequency.

Worked example 8.6

Construct a formal definition of each of the following processes, indicating appropriate pre- and post-conditions:
(a) counting the number of words in a document;
(b) finding the square of a number;
(c) library enquiry system that gives the location number of a particular title;
(d) listening to the radio.

Solution:
(a) wordcount: <string> $\mapsto$ wordcount(<string>) $= n$

where

<string> $\in$ {s: s is a string of characters} and $n \in \mathcal{N}$

(b) square: $x \mapsto$ square$(x) = x^2$

where

$x \in \mathcal{R}$ and $x^2 \in \mathcal{R}$

(c) titlelocation: <string> $\mapsto$ titlelocation(<string>) $=$ code

where

<string> $\in$ {rec: rec is a record in the library database}

and

code ∈ {ref: ref is a reference location code}

(d) radio: channel ↦ radio(channel) = programme

where

channel ∈ {freq: freq is a radio frequency}

and

program ∈ {p: p is a radio programme}

Worked example 8.7

Construct the graphs of each of the following functions:
(a) square: $x \mapsto$ square$(x) = x^2$

where

$x \in \{-2, -1, 0, 1, 2\}$

(b) radio: channel ↦ radio(channel) = programme

where

channel ∈ {88.9 MHz, 90.2 MHz, 92.4 MHz, 96.7 MHz}

and

programme ∈ {Radio 1, Radio 2, Radio 3, Radio 4}

Solution:
(a) {(−2, 4), (−1, 1), (0, 0), (1, 1), (2, 4)}
(b) {(88.9 MHz, Radio 2), (90.2 MHz, Radio 3), (92.4 MHz, Radio 4), (96.7 MHz, Radio 1)}

Exercises

8.3 Describe in words the action of each of the following functions:
  (a) measuring the length of a journey;
  (b) finding the cube root of a number;
  (c) selecting a holiday from a brochure;
  (d) spellchecking a document.

8.4 Describe the domain and range of each of the following functions:
  (a) measuring the length of a journey;
  (b) finding the cube root of a number;
  (c) selecting a holiday from a brochure;
  (d) spellchecking a document.

8.5 Which of the following are total functions? Give appropriate pre-conditions for those that are partial functions to convert them into total functions:

(a) measuring the length of a journey;
(b) finding the cube root of a number;
(c) selecting a holiday from a brochure;
(d) spellchecking a document.

8.6 Construct a formal definition of each of the following processes, indicating appropriate pre- and post-conditions:

(a) measuring the length of a journey;
(b) finding the cube root of a number;
(c) selecting a holiday from a brochure;
(d) spellchecking a document.

8.7 Construct the graphs of each of the following functions:

(a) finding the cube root of a number where the domain is restricted to

$$\{-27, -8, 0, 8, 27\}$$

(b) selecting a holiday from a brochure.

Properties of mappings

Types of mapping
There are three distinct types of mapping.

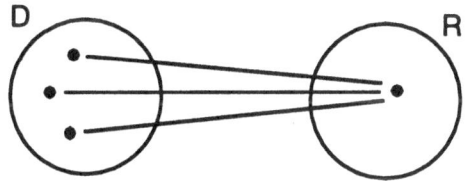

Figure 8.7

The first map (Figure 8.7) is called a **many-to-one** map because many points in the domain are mapped to a single range point. Typical of such a map is the function square that squares the input:

square: $x \mapsto$ square$(x) = x^2$

where

x, square$(x) \in \mathcal{R}$

For example

The use of the word **map** here differs from its use cartographically. Here the map is that process that links a domain value with a range value. In an atlas a map is the range.

square(2) = 4
square(−2) = 4

so that both the domain points 2 and −2 map to the single range point 4.

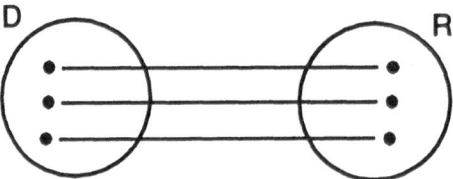

Figure 8.8

The second map (Figure 8.8) is called a **one-to-one** map because single domain points are mapped to single range points and no two or more domain points map to the same range point. Typical of such a map is the function recip that outputs the reciprocal of the input:

recip: $x \mapsto \text{recip}(x) = 1/x$

where

$x \in \mathcal{R} \wedge x \neq 0, \text{recip}(x) \in \mathcal{R}$

Since every number has a unique reciprocal this is a one-to-one function.

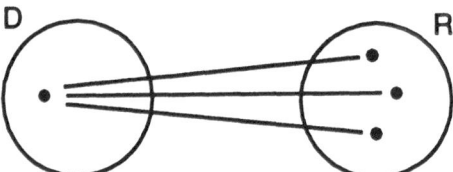

Figure 8.9

The third map (Figure 8.9) is called a **one-to-many** map because there are domain points that individually map to more than one range point. Typical of such a map is the process birthday that gives the name of an individual whose birthday is on a given date:

birthday: <date> $\mapsto$ birthday(<date>)

where

<date> $\in \{d: d$ is a valid date$\}$

and

birthday(<date>) $\in$ {p: p is a person}

Obviously many people share the same birthday.

Only the first two types of mappings describe functions. The third mapping, a one-to-many mapping, describes a different entity which we shall discuss later.

One-to-one mapping

Consider the mapping, called map, where

map: $x \mapsto \text{map}(x)$

where

$x \in D$

and

$\text{map}(x) \in R$

This mapping represents a function if two different range points are associated with two different domain points. That is, the condition for a mapping to represent a function can be given as

If $\text{map}(x) \neq \text{map}(y)$ then $x \neq y$

where

$x, y \in D$

and

$\text{map}(x), \text{map}(y) \in R$

The contrapositive of this statement is also valid, namely

If $x = y$ then $\text{map}(x) = \text{map}(y)$

where

$x, y \in D$

and

$\text{map}(x), \text{map}(y) \in R$

Clearly, if the same input is entered on two different occasions then the same output appears – the function must be consistent.

The converse of the original statement is not necessarily true:

If $x \neq y$ then $\text{map}(x) \neq \text{map}(y)$

If this converse is true then the function is said to be **one-to-one** and different inputs correspond to different outputs. If this converse is false and the mapping is a function then the function is **many-to-one** – more than one input may correspond to the same output.

The inverse of the original statement is the contrapositive of the converse:

If map(x) = map(y) then $x = y$

Again, this is only true if the function is one-to-one.

Conclusion: A mapping is a process whereby an element of the domain of the mapping is linked to an element of the range of the mapping. There are three types of mapping: one-to-one, many-to-one and one-to-many. Only the first two describe functions.

Worked example 8.8

Which of the following mappings represent a function:
(a) date: day $\mapsto$ date(day)

where

day $\in$ {days of the week}

and

date(day) $\in$ {valid date of the year}
(b) cuberoot: $x \mapsto$ cuberoot(x) = $x^{1/3}$

where

$x, x^{1/3} \in \mathcal{R}$
(c) balance: account $\mapsto$ balance(account) = amount

where

account $\in$ {valid bank account}

and

amount $\in$ {a: a is a number to two decimal places}
(d) account: balance $\mapsto$ account(balance) = code

where

balance $\in$ {b: b is a number to two decimal places}

and

code $\in$ {valid bank account code}

Solution:
(a) This is not a function because the same day, for example Monday, occurs on 52 different dates of the year.
(b) This is a function because every real number has a unique cube root.

(c) This is a function because every account has a single balance.

(d) This is a not function because more than one account may have the same balance.

Worked example 8.9

Which of the following functions are one-to-one and which are many-to-one?

(a) cuberoot: $x \mapsto$ cuberoot$(x) = x^{1/3}$

where

$x, x^{1/3} \in \mathcal{R}$

(b) balance: account $\mapsto$ balance(account) = amount

where

account $\in$ {valid bank account}

and

amount $\in$ {a: a is a number to two decimal places}

Solution:

(a) This function is one-to-one because each real number has a single, unique cube root.

(b) This function is many-to-one because two or more accounts may have the same balance.

Exercises

8.8 Which of the following mappings represent a function?

(a) sum: \<list\> $\mapsto$ sum(\<list\>)

where

\<list\> is a list of numbers

and

sum(\<list\>) is the arithmetic sum of the numbers in the list

(b) item: invent $\mapsto$ item(invent)

where

invent is a natural number

and

item(invent) $\in$ {items in stock}

(c) wordcount: document $\mapsto$ wordcount(document) = n

where

document is a list of words

and

$n \in \mathcal{N}$

(d)   distance: destination $\mapsto$ distance(destination)

where

destination $\in$ {towns in West Yorkshire}

and

distance(destination) $\in$ {distance between Huddersfield and destination}

8.9 Which of the following functions are one-to-one and which are many-to-one?

(a)   sum: <list> $\mapsto$ sum(<list>)

where

<list> is a list of numbers

and

sum(<list>) is the arithmetic sum of the numbers in the list

(b)   wordcount: document $\mapsto$ wordcount(document) $= n$

where

document is a list of words and $n \in \mathcal{N}$

Composition of functions

When output becomes input

We have discussed how a computer program can be described as a function. In practice, this may be a wild oversimplification because in even a quite simple program it may not be best to describe it as a single function. As an illustration consider how, with a hand calculator, you would compute

$$(3.4)^2 + 2.6$$

First you perform the squaring:

Enter the number 3.4
Press the square function key the display of the result 11.56

This is one function. Next you perform the addition

Press the + key
Enter the number 2.6
Press the = key to produce the display of the result 14.16

This is another function.

This simple arithmetic process has involved two functions operated in sequence where the output from the first function, namely 11.56, becomes the input for the second function. Functions that are linked in this way are said to be **compositions** and can be represented using a sequentially linked system diagram (Figure 8.10).

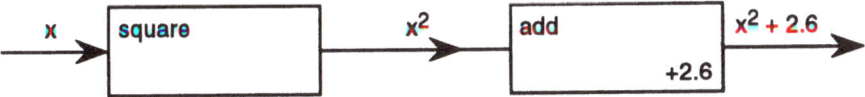

Figure 8.10

If we give the signatures of these two functions as:

square: $x \mapsto$ square$(x) = x^2$

add: $x \mapsto$ add$(x) = x + 2.6$

we define the signature of their composition as:

add $\circ$ square: $x \mapsto$ add(square$[x]) = x^2 + 2.6$

where the operation of composition is represented by $\circ$. Notice that the signature of the composition builds up to the left, whereas the diagram of the composition builds up to the right.

Conclusion: If the output from one function forms the input to a subsequent function the two functions are said to be composed together. Many involved functions can be decomposed into a composition of simpler functions.

Worked example 8.10

Given the functions

add3: $x \mapsto$ add3$(x) = x + 3, x \in \mathcal{R}$
mult5: $x \mapsto$ mult5$(x) = 5x, x \in \mathcal{R}$
cube: $x \mapsto$ cube$(x) = x^3, x \in \mathcal{R}$

find the output from each of the following compositions:
(a)   add3 $\circ$ cube
(b)   cube $\circ$ add3
(c)   mult5 $\circ$ add3 $\circ$ cube
(d)   add3 $\circ$ add3 $\circ$ cube $\circ$ mult5

Solution:
(a)   add3 $\circ$ cube: $x \mapsto$ add3$[$cube$(x)] =$ add3$[x^3] = x^3 + 3$

(b) cube ∘ add3: $x \mapsto$ cube[add3(x)] = cube[$x + 3$] = $(x + 3)^3$

(c) mult5 ∘ add3 ∘ cube: $x \mapsto$ mult5{add3[cube(x)]}

$$= \text{mult5}\{\text{add3}[x^3]\}$$
$$= 5(x^3 + 3)$$

(d) add3 ∘ add3 ∘ cube ∘ mult5: $x \mapsto$ add3{add3[cube(mult5{x})]}

$$= \text{add3}\{\text{add3}[125x^3]\}$$
$$= \text{add3}\{125x^3 + 3\}$$
$$= 125x^3 + 6$$

Worked example 8.11

Break each of the following function compositions into their component parts:

(a) comp1: $x \mapsto$ comp1$(x) = 7(x^2 + 6)$, $x \in \mathcal{R}$

(b) comp2: $x \mapsto$ comp2$(x) = (7x)^2 + 6$, $x \in \mathcal{R}$

(c) comp3: $x \mapsto$ comp3$(x) = (7x + 6)^2$, $x \in \mathcal{R}$

(d) comp4: $x \mapsto$ comp4$(x) = [7(7x + 6)]^2 + 6$, $x \in \mathcal{R}$

Solution:

(a) comp1 = mult7 ∘ add6 ∘ square

(b) comp2 = add6 ∘ square ∘ mult7

(c) comp1 = square ∘ add6 ∘ mult7

(d) comp1 = add6 ∘ square ∘ mult7 ∘ add6 ∘ mult7

Exercises

8.10 Given the functions

sub4: $x \mapsto$ sub4$(x) = x - 4$, $x \in \mathcal{R}$

div8: $x \mapsto$ div8$(x) = x/8$, $x \in \mathcal{R}$

cuberoot: $x \mapsto$ cuberoot$(x) = x^{1/3}$, $x \in \mathcal{R}$

find the output from each of the following compositions:

(a) sub4 ∘ div8

(b) div8 ∘ sub4

(c) div8 ∘ sub4 ∘ cuberoot

(d) div8 ∘ div8 ∘ cuberoot ∘ sub4

8.11 Break each of the following function compositions into their component parts:

(a) comp5: $x \mapsto$ comp5$(x) = 2(x^5 - 9)$, $x \in \mathcal{R}$

(b) comp6: $x \mapsto$ comp6$(x) = (2x)^3 - 9$, $x \in \mathcal{R}$

(c) comp7: $x \mapsto$ comp7$(x) = (2x - 9)^3$, $x \in \mathcal{R}$

(d) comp8: $x \mapsto$ comp8$(x) = [2(2x - 9)]^3 - 9$, $x \in \mathcal{R}$

Inverse functions

Reversing processes

Many of the processes that we have considered are capable of having their reverse process described. For example, the process that adds 5 to an input has the obvious reverse process that subtracts 5 from an input (Figure 8.11).

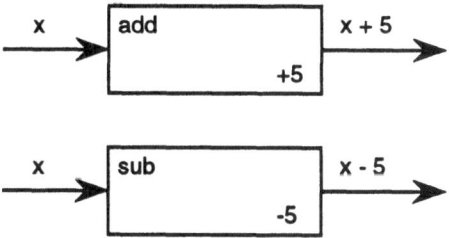

Figure 8.11

Such reverse processes are referred to as **inverse** processes. If the process is a function then the reverse process is referred to as the **inverse of** the function. If the inverse of the function is itself a function then it is called the **inverse function**. For example, the function that subtracts 5 from an input is the inverse of the function that adds 5 to an input. The inverse of the function is also a function and is, therefore, the inverse function. On the other hand, the function that counts the number of vowels in a string inverts to the process where an input number is converted to a string containing that number of vowels. This reverse process, the inverse of the function, is clearly not a function – there are countless strings each containing the same number of vowels – so it cannot be described as the inverse function. The distinction depends on the type of mapping that represents the original function.

The inverse of a one-to-one function is the inverse function (Figure 8.12)

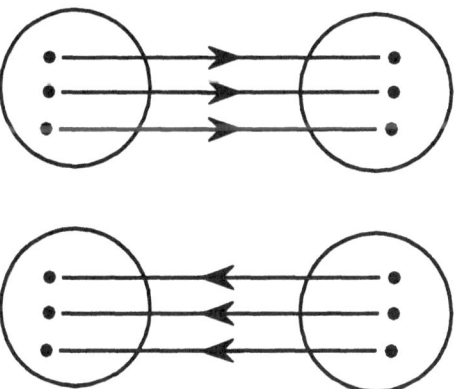

Figure 8.12

The inverse of a many-to-one function is not a function (Figure 8.13)

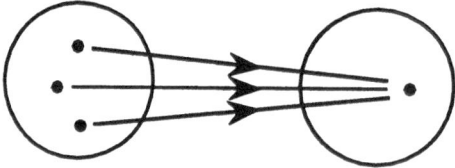

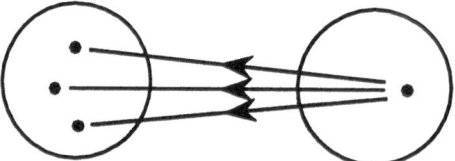

Figure 8.13

As an example of the application of finding the inverse of a one-to-one function consider the process of converting a decimal natural number to its binary counterpart. This process can be thought of as a function decbin, where

decbin: $x \mapsto$ decbin(x)

where

$x \in \mathcal{N}$

and

decbin(x) $\in$ {b: b is a binary number}

For example

decbin(23) = 10111

This process can be reversed by converting a binary number into its decimal form. This reversed function can also be thought of as a function bindec where:

bindec: $x \mapsto$ bindec(x)

where

$x \in$ {x: x is a binary number}

and

bindec(x) $\in \mathcal{N}$

For example

bindec(10111) = 23

The function bindec, which is the reverse process of decbin, is called the **inverse** function of decbin. Similarly, decbin is the inverse function of bindec – they are **mutual inverses.**

Every function has an inverse which may or may not be a function itself. If it is a function then it is called the inverse function. For example, consider the many-to-one function ageof where

ageof: <name> $\mapsto$ ageof(<name>)

where

<name> $\in$ {*n*: *n* is a person's name}

and

ageof(<name>) $\in \mathcal{R}$

This function accepts a person's name <name> and outputs that person's age. The reverse of this process outputs a person's name for a given input of age; this is the inverse of the function ageof. However, this reverse process is represented by a one-to-many mapping because to any given age there will correspond more than one person. Consequently, it is not a function and, therefore, while it is the inverse of the function it is not the inverse function. The only time the inverse of a function is the inverse function is if the original function is a one-to-one function.

Notation
We called bindec the inverse function of the function decbin because the juxtaposition of the component parts of the two names was appropriate. Most times, however, this is not the case. For example, the function

name: $x \mapsto$ name(x) = NAME

where

$x \in \{1, 2, 3, 4\}$

and

NAME $\in$ {Adams, Bassett, Carter, Davis}

is a one-to-one function and will possess an inverse that is the inverse function. To give the inverse function a name that indicates that it is an inverse function we use the notation

name^{-1}: NAME $\mapsto$ name^{-1}(NAME) = x

where

NAME ∈ {Adams, Bassett, Carter, Davis}

and

$x \in \{1, 2, 3, 4\}$

Take especial care with this notation. The -1 in the signature of the inverse function does not mean that the inverse function is in any way related to a reciprocal. It is just bad notation which is, unfortunately, commonly accepted. A better notation could make use of the prefix inv:

invname: NAME ↦ invname(NAME) = x

where

NAME ∈ {Adams, Bassett, Carter, Davis}

and

$x \in \{1, 2, 3, 4\}$

Conclusion: If the input to and output from a function is reversed the resulting process that is achieved is called the inverse of the function. If the inverse of a function is itself a function then it is called the inverse function.

Worked example 8.12

Describe the action of the inverse of each of the following functions:
(a) dechex: $h \mapsto$ dechex(h)

where

$h \in$ {hexadecimal number}

and

dechex(x) ∈ $\mathcal{N}$

(b) occupy: address ↦ occupy(address) = n

where

$n \in \mathcal{N}$

and

address ∈ {valid address}

(c) surname: recordnumber ↦ surname(recordnumber) = sname

where

recordnumber ∈ $\mathcal{N}$

and

sname ∈ {surnames}

(d) add5: $x \mapsto \text{add5}(x) = x + 5$

where

$x \in \mathcal{R}$

Solution:
(a) dechex converts a number in hexadecimal form to its equivalent decimal form. The action of the inverse is to convert a number in decimal form into its equivalent hexadecimal form. This inverse process is a function and is, therefore, the inverse function.
(b) occupy returns the number of occupants living at a given address. The inverse process will give the address(es) that correspond to a given number of occupants. Because two or more addresses will have the same number of occupants the inverse process is not a function.
(c) surname returns the surname located on a record with a given number. The reverse process will give the record number(s) that corresponds to a given surname. This inverse process is not a function as more than one record may have the same surname.
(d) add5 adds 5 to the input. The reverse process will subtract 5 from the input. The inverse process is a function.

Worked example 8.13

Suggest an inverse function for each of the functions of the previous question.

Solution:
(a) A suitable inverse function would be hexdec, which converts a decimal number into its hexadecimal representation.
(b) A suitable inverse process would be occnumber, which returns the address shared by a given number of people. Note that this may not be a function because the different addresses may be shared by the same number of people.
(c) A suitable inverse process would be recnumber, which returns the record number of the record which contains a given surname. Note that this may not be a function because the same surname may occur on more than one record.
(d) A suitable inverse function would be sub5, which subtracts 5 from the input.

Worked example 8.14

Find the inverse of each of the following compositions:
(a) decbin ∘ binhex ∘ hexdec
(b) add5 ∘ mult6 ∘ cube
(c) sub3 ∘ add3

Solution:
(a) dechex ∘ hexbin ∘ bindec
(b) cuberoot ∘ div6 ∘ sub5
(c) sub3 ∘ add3

Exercises

8.12 Describe the action of the inverse of each of the following functions:
   (a) binhex: $h \mapsto$ binhex(h)

   where

   $h \in$ {hexadecimal number}

   and

   binhex(h) $\in$ {b: b is a binary number} (the binary form of h)

   (b) number: <department> $\mapsto$ number(<department>) $= n$

   where

   department $\in$ {department name}

   and

   $n \in \mathcal{N}$ (the number working in the department)

   (c) salary: recordnumber $\mapsto$ salary(recordnumber) = amount

   where

   recordnumber $\in \mathcal{N}$

   and

   amount $\in$ {number to two decimal places}

   (d) div8: $x \mapsto$ div8(x) = $x/8$

   where

   $x \in \mathcal{R}$

8.13 Suggest an inverse function for each of the functions of the previous question.

8.14 Find the inverse of each of the following compositions:
   (a) binhex ∘ hexdec ∘ dechex
   (b) div6 ∘ cuberoot ∘ square
   (c) sub8 ∘ mult4

Functions and relations

The graphs of all the functions that we have considered so far have been subsets of the Cartesian product of the domain and the range of the function

$G \in D \times R$

In the previous chapter we defined a relation as a subset of a Cartesian product so it would appear that all graphs of functions are relations. Indeed, that is the case though the converse is not true: not all relations are graphs of functions.

We have seen that a function can be expressed as a one-to-one or many-to-one mapping because to each input value to a function there is associated just one output value. The third type of mapping, the one-to-many mapping, does not represent a function but it does represent a relation. Consequently the set of graphs of functions forms a subset of the set of relations.

Conclusion: The graph of a function is a relation but a relation is only the graph of a function if the relation can be represented by a one-to-one or many-to-one mapping.

Worked example 8.15

Which of the following relations are also graphs of functions?
(a) $R = \{(x, y): x, y \in \mathcal{R} \wedge x < y\}$
(b) $R = \{(x, y): x, y \in \mathcal{N} \wedge x = y + 1\}$
(c) $\{(1, 2), (2, 3), (3, 2), (2, 1), (1, 1), (2, 2), (3, 3)\}$
(d) $\{(1, 2), (2, 1), (3, 4), (4, 3), (5, 6), (6, 5)\}$
(e) $\{(UK, RWB), (F, RWB), (USA, RWB), (H, RW), (C, RW)\}$

Solution:
(a) This relation is not the graph of a function because to each first element of an ordered pair there is an infinity of corresponding second elements. For example, (0, 1) and (0, 2) both satisfy the prescription $x < y$, thereby violating the requirement for a function to be single valued.
(b) This relation is the graph of a function because to each first element of the ordered pair (x, y) there is a unique second element.
(c) The existence of (1, 1) and (1, 2) ensures that this relation is not the graph of a function.
(d) To each first element of the ordered pairs of this relation there corresponds a unique second element. Consequently this is the graph of a function.
(e) To each first element of the ordered pairs of this relation there corresponds a unique second element. Consequently this is the graph of a function.

Worked example 8.16

Given the following three relations

R = {(1, 2), (2, 3), (3, 2), (2, 1), (1, 1), (2, 2), (3, 3)}
S = {(1, 2), (2, 1), (3, 4), (4, 3), (5, 6), (6, 5)}
T = {(x, y) : x, y ∈ {0, 1, 2, 3, 4, 5, 6} ∧ y = x − 1}

find

(a) The composition R ∘ S
(b) The inverse of T
(c) The inverse of R ∘ S

Solution:

(a) The composition R ∘ S is formed by matching the second elements of S with the first elements of R to form an ordered pair of the composition. For example, the ordered pair (1, 2) of the relatioñ S has 2 as the second element. This is matched with those ordered pairs of R which have 2 as their first element, namely

(2, 3), (2, 1) and (2, 2)

The composition of these ordered pairs gives

(1, 2) ∘ (2, 3) = (1, 3)
(1, 2) ∘ (2, 1) = (1, 1) and
(1, 2) ∘ (2, 2) = (1, 2)

The composition R ∘ S is therefore

{(1, 3), (1, 1), (1, 2), (2, 2), (2, 1), (4, 2), (4, 3)}

(b) The inverse of T is obtained by interchanging the role of input and output. This can be achieved by interchanging the symbols in the general ordered pair to give

$T^{-1} = \{(y, x): x, y ∈ \{0, 1, 2, 3, 4, 5, 6\} ∧ y = x − 1\}$

If we now relabel the elements of each ordered pair to conform with the accepted notation that x represents the first element and y the second then we must relabel the x's and the y's in the rest of the definition of the inverse. This gives

$T^{-1} = \{(x, y): x, y ∈ \{0, 1, 2, 3, 4, 5, 6\} ∧ x = y − 1\}$

The prescription $x = y − 1$ can be written as:

$y = x + 1$

The inverse of the process of subtracting 1 from an input is the addition of 1 to the input.

(c) From part (a) we have that

R ∘ S = {(1, 3), (1, 1), (1, 2), (2, 2), (2, 1), (4, 2), (4, 3)}

The inverse $(R ∘ S)^{-1}$ is obtained by interchanging elements in each ordered pair:

$(R ∘ S)^{-1} = \{(3, 1), (1, 1), (2, 1), (2, 2), (1, 2), (2, 4), (3, 4)\}$

Exercises

8.15 Which of the following relations are also graphs of functions:

(a) $R = \{(x, y): x, y \in \mathcal{R} \wedge x = y^2\}$
(b) $R = \{(x, y): x, y \in \mathcal{N} \wedge x^2 = y - 3\}$
(c) $\{(a, b), (b, c), (c, b), (b, a), (a, a), (b, b), (c, c)\}$
(d) $\{(a, b), (b, a), (c, d), (d, c), (e, f), (f, e)\}$
(e) $\{(Ford, red), (Renault, red), (Ford, blue), (Volvo, white), (Chrysler, black)\}$

8.16 Given the following three relations:

$R = \{(a, b), (b, c), (c, b), (b, a), (a, a), (b, b), (c, c)\}$
$S = \{(a, b), (b, a), (c, d), (d, c), (e, f), (f, e)\}$
$T = \{(x, y): x, y \in \{1, 2, 3, 4, 5, 6\} \wedge x = y - 2\}$

Find

(a) The composition $R \circ S$
(b) The inverse of T
(c) The inverse of $R \circ S$

Recursive processes

Sequences of terms

The contents of a list form a **sequence of terms**. For example, the list

<red, orange, yellow, green, blue, indigo, violet>

contains the sequence of terms

red, orange, yellow, green, blue, indigo, violet

where each term of the sequence is a colour of the rainbow. Notice that this sequence is different from the sequence

yellow, green, red, orange, blue, indigo, violet

because the individual terms of the sequence are in a different order and, therefore, form the contents of a different list.

A sequence of a finite number of terms is called a finite sequence and a sequence of an infinite number of terms is called an infinite sequence. If the terms of an infinite sequence can be identified precisely then each term can be associated with a natural number:

the first term is associated with   1
the second term is associated with   2
the third term is associated with   3
...   .
the nth term is associated with n

Such sequences are said to be **countable** even if the sequence consists of an infinite number of terms. A sequence that is not countable is said to be **uncountable**. For example, consider the set of all numbers between 0 and 1. Each number can be represented as an infinite decimal:

$$a_1 = 0.p_1p_2p_3p_4p_5\ldots$$
$$a_2 = 0.q_1q_2q_3q_4q_5\ldots$$
$$a_3 = 0.r_1r_2r_3r_4r_5\ldots$$
$$\ldots$$

This sequence gives the appearance of being countable because each decimal number a_i, $i \in \mathcal{N}$ is distinguished from the next decimal number by the suffix i, which is a natural number. In fact this appearance of countability is an illusion because the numbers cannot be so ordered. This statement can be proved by demonstrating that even though we assume the list to be complete it is always possible to define a number that is not already on the list. We shall define the infinite decimal b_1 as follows:

$$b_1 = 0.x_1x_2x_3x_4x_5\ldots$$

If $p_1 = 0$ then $x_1 = 1$ otherwise $x_1 = 0$
If $q_2 = 0$ then $x_2 = 1$ otherwise $x_2 = 0$
If $r_3 = 0$ then $x_3 = 1$ otherwise $x_3 = 0$
$\ldots$

In this way we guarantee that b_1 is an infinite decimal that is different from every other decimal in the list, so demonstrating by *reductio ad absurdum* that when the list is assumed to be fully constructed it is always possible to construct another member.

Recognizing sequences
All the sequences that we shall deal with will be countable sequences and, moreover, are capable of having their terms generated by a rule. For example, the sequence whose first five terms are

1, 3, 9, 27, 81,…

can be generated by the rule that each term is obtained by multiplying the previous term by 3. Clearly, the next three terms of this sequence are then

243, 729, 2187

Finding the rule is not always a simple process.

Conclusion: The contents of a list, in the order in which they appear in the list form a sequence of terms. If each term of the sequence can be identified

as distinct from any other term in the sequence the sequence is said to be countable.

Worked example 8.17

Which of the following sequences are countable?
(a) 1, 1.1, 1.11, 1.111, 1.1111,... *ad infinitum*;
(b) The sequences of rational numbers between 0 and 1.

Solution:
(a) There is clearly an infinite number of terms in this sequence but each term can be distinguished from any other term. For example, count the number of 1's in each term and list them as follows:

Term:	1	1.1	1.11	1.111	1.1111	...
Number of 1's	1	2	3	4	5	...

Here we see that each term is uniquely associated with one of the naturals thereby demonstrating that the sequence is countable.
(b) Again, this sequence has an infinite number of terms but each term can be distinguished from any other term of the sequence thereby demonstrating that they are countable. For example, list the rational as follows:

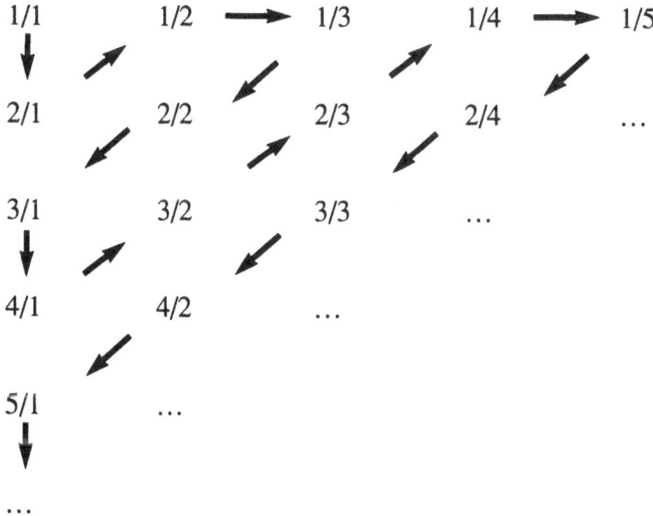

By listing and moving through the list in the direction of the arrows in the way illustrated we ensure that every rational is uniquely identified, therefore demonstrating that the rational numbers are countable.

Worked example 8.18

Find the next three terms in each of the following sequences:
(a) 7, 10, 13, 16,...
(b) 1, 5, 25, 125,...
(c) 1/2, $-1/4$, 1/8, $-1/16$,...
(d) 0, 0.1, 0.12, 0.123, 0.1234,...

Solution:
(a) Each term is obtained from the previous one by adding 3. Thus the next three terms are 19, 22, 25.
(b) Each term is obtained from the previous one by multiplying by 5. Thus the next three terms are 625, 3125, 15625.
(c) Each term is obtained from the previous one by multiplying by $-1/2$. Thus the next three terms are 1/32, $-1/64$, 1/128.
(d) Each term is obtained from the previous one by placing the next natural number in the next decimal place. Thus the next three terms are 0.12345, 0.123456, 0.1234567.

Exercises

8.17 Which of the following sequences are countable?
   (a) All the fine gradations of colours in a rainbow.
   (b) 2, 4, 6, 8, 10,...

8.18 Find the next three terms in each of the following sequences:
   (a) 3, -1, -5, -9,...
   (b) 2, 12, 72, 432,...
   (c) $-1/3$, 2/9, $-3/27$, 4/81,...
   (d) 1.01, 2.002, 3.0003, 4.00004,...

Sequences as functions

Because each term in a countable sequence of terms can be associated with a unique natural number by virtue of their order of listing, the individual terms can be identified as the output from a function. For example, the function square whose input values are restricted to the natural numbers has the following signature:

square: $n \mapsto \text{square}(n) = n^2$

where

$n, \text{square}(n) \in \mathcal{N}$

This function has outputs

$$0, 1, 4, 9, 16, 25, 36, 49,\ldots$$

corresponding to inputs

$$1, 2, 3, 4, 5, 6, 7,\ldots$$

Sequences abound and are of many and varied types. Two types of particular interest are arithmetic and geometric sequences.

Arithmetic and geometric sequences
An arithmetic sequence is any sequence of the form

$$\text{arith}: n \mapsto \text{arith}(n) = a + nd$$

where

$$n \in \mathcal{N}$$

and

$$a, d, \text{arith}(n) \in \mathcal{R}$$

The progressive output from this sequence is

$$a, a + d, a + 2d, a + 3d,\ldots$$

where a is called the first term and d is called common difference, being the difference between any term and its predecessor. For example, the sequence

$$\text{even}: n \mapsto \text{even}(n) = 2n$$

where

$$n \in \mathcal{N}$$

is an arithmetic sequence whose successive terms are

$$2, 4, 6, 8, 10,\ldots$$

In this arithmetic sequence the first term is 0 and the common difference is 2. A geometric sequence is any sequence of the form

$$\text{geom}: n \mapsto \text{geom}(n) = Ar^n$$

where

$n \in \mathcal{N}$ and $A, r, \text{geom}(n) \in \mathcal{R}$

The progressive output from this sequence is

$A, Ar, Ar^2, Ar^3, \ldots$

where A is called the first term and r is called common ratio, being the ratio between any term to its predecessor. For example, the sequence

three: $n \mapsto \text{three}(n) = 3^n$

where

$n \in \mathcal{N}$

is a geometric sequence whose successive terms are

$1, 3, 9, 27, 81, 243, \ldots$

In this geometric sequence the first term is 1 and the common ratio is 3.

Conclusion: The terms of a sequence can be generated by a function and typical of such generation produces the arithmetic and geometric sequences.

Worked example 8.19

Write down the first four terms of each of the following sequences:
(a) $\text{arith}(n) = 5 - 2n$, $n = 0, 1, 2, \ldots$
(b) $\text{geom}(n) = 3(0.2)^n$, $n = 0, 1, 2, \ldots$

Solution:
(a)   By simple application of the formula we find

$\text{arith}(0) = 5 - 2 \times 0 = 5$
$\text{arith}(1) = 5 - 2 \times 1 = 3$
$\text{arith}(2) = 5 - 2 \times 2 = 1$
$\text{arith}(3) = 5 - 2 \times 3 = -1$

(b)   By simple application of the formula we find

$\text{geom}(0) = 3(0.2)^0 = 3$
$\text{geom}(1) = 3(0.2)^1 = 0.6$
$\text{geom}(2) = 3(0.2)^2 = 0.12$
$\text{geom}(3) = 3(0.2)^3 = 0.024$

Worked example 8.20

The sum of the first n terms of the arithmetic sequence arith$(n) = a + nd$ is given as

arithsum$(n) = [n/2](2a + [n - 1]d)$

Show that the sum of the first n integers is given by

$[n/2](n + 1)$

Solution: The sequence of the first n integers

$1, 2, 3,..., n$

is an arithmetic sequence with first term 1 and common difference 1. Consequently the sum of the first n terms is

arithsum$(n) = [n/2](2 \times 1 + [n - 1] \times 1)$
$\qquad\qquad = [n/2](2 + n - 1)$
$\qquad\qquad = [n/2](n + 1)$

Worked example 8.21

The sum of the first n terms of the geometric sequence geom$(n) = Ar^n$ is given as

geomsum$(n) = A(1 - r^n)/(1 - r)$

Show that the sum of the first n terms of the sequence

$1, 1/2, 1/4, 1/8,...$

is given as

$2(1 - 1/2^n)$

Solution: In this geometric sequence the first term is 1 and the common ratio is 1/2. The sum of the first n terms is, therefore

$1(1 - [1/2]^n)/(1 - 1/2) = (1 - 1/2^n)/(1/2)$
$\qquad\qquad\qquad\qquad = 2(1 - 1/2^n)$

Worked example 8.22

The third, fifth and eighth terms of an arithmetic sequence form the first three terms of a geometric sequence. Find the geometric sequence if the first term of the arithmetic sequence is 1.

Solution: Denote the general term of the arithmetic sequence by arith(n) where

arith(n) = $1 + nd$

Denote the general term of the geometric sequence by geom(n) where

geom(n) = Ar^n

The problem then states that

arith(3) = $1 + 3d$ = geom(0) = $Ar^0 = A$,
arith(5) = $1 + 5d$ = geom(1) = $Ar^1 = Ar$, and
arith(8) = $1 + 8d$ = geom(2) = $Ar^2 = Ar^2$

Taking ratios we see that

$(1 + 3d)/(1 + 5d) = A/Ar = Ar/Ar^2 = (1 + 5d)/(1 + 8d)$

Thus

$(1 + 3d)/(1 + 5d) = (1 + 5d)/(1 + 8d)$

That is

$(1 + 3d)(1 + 8d) = (1 + 5d)(1 + 5d)$

which means that

$1^2 + 11d + 24d^2 = 1^2 + 10d + 25d^2$

Hence

$d - d^2 = d(1 - d) = 0$

Consequently, $d = 1$ or $d = 0$. Since $A = 1 + 3d$ we see that $A = 1$ if $d = 0$ and $A = 4$ if $d = 1$. Also $Ar = 1 + 5d$. If $d = 0$ then $A = 1$, $Ar = 1$ so that $r = 1$. This gives the geometric sequences as

geom(n) = 1

Again, $Ar = 1 + 5d$. If $d = 1$ then $A = 4$, $Ar = 6$ so that $r = 3/2$. This gives the geometric sequences as

geom(n) = $4(3/2)^n$

Exercises

8.19 Write down the first four terms of each of the following sequences:

(a) arith(n) = $6n + 3$, $n = 0, 1, 2,...$
(b) geom(n) = $4(-1)^n$, $n = 0, 1, 2,...$

8.20 The sum of the first n terms of the arithmetic sequence arith(n) = $a + nd$ is given as

arithsum(n) = $[n/2](2a + [n - 1]d)$

Show that the sum of the first n odd integers is given as

n^2

8.21 The sum of the first n terms of the geometric sequence geom(n) = Ar^n is given as

geomsum(n) = $A(1 - r^n)/(1 - r)$

Show that the sum of the first n terms of the sequence

$1, -1/2, 1/4, -1/8,...$

is given as

$2(1 - [-1/2]^n)/3$

8.22 The third, fifth and eighth terms of an arithmetic sequence form the first three terms of a geometric sequence. The first term of the arithmetic sequence is 4. Find the geometric sequence.

Recursive sequences

We use mathematics to model the world as we perceive it with the express purpose of being able to predict its behaviour. Many situations exist in which the observed behaviour of some physical process can be modelled by the successive terms of a sequence, and it is a natural reaction to model that physical process by using a function. However, while the output from the function may simulate the observed outcome of the physical process, if we cannot describe the action of the function explicitly we are no nearer to explaining the action of the physical world. This can make prediction difficult if not impossible to achieve, thereby defeating our purpose. Situations do exist, however, where the natural world produces effects that follow a **pattern**, and even when we cannot quantify the processes that produce the pattern we can at least describe the pattern and thereby predict the future. Take, for example, the case of the multiple reflections of a ray of light striking two sheets of glass that are in flat contact (Figure 8.14).

Here we see that if we correlate the number of internal reflections an incident ray of light can make with the number of different ways it can make those reflections we obtain the following sequence:

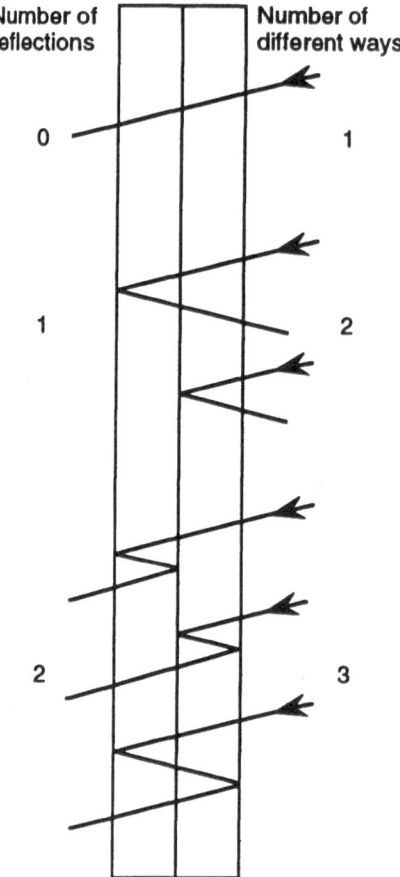

Figure 8.14

Reflections	0	1	2	3	4	5	6	7
Ways	1	2	3	5	8	13	21	34

Here we have found the outcome of a physical behaviour that can be simulated by the successive terms of a sequence, but how are we to describe the action of the function that produces the successive output values for corresponding input values? The answer to this question is not immediately obvious — indeed, there may not even be an answer. However, what we do notice is that each term in the sequence is the sum of its two predecessors:

$1 + 2 = 3$
$2 + 3 = 5$
$3 + 5 = 8$
$5 + 8 = 13$

. . .

This is the pattern and, whilst we cannot quantify the action of the function that produces the pattern, we can quantify the pattern by using the function fib, where an output is given as the sum of the two previous outputs:

$$\text{fib}: n \mapsto \text{fib}(n + 2) = \text{fib}(n) + \text{fib}(n + 1)$$

where

$$n, \text{fib}(n) \in N$$

The function fib gives rise to a sequence known as a **Fibonacci** sequence, named after its discoverer. It is an important sequence because it appears in the natural world in many different situations: pairs of breeding rabbits, the spiral arrangement of the seeds on the head of a sunflower, the structure of the nautilus shell and, not least, the human experience of beauty in relation to the **golden mean**.

You will notice that we have not quantified the action of fib but merely defined it in terms of itself – we have defined it recursively, just as we did when we considered the definition of the arithmetic operations in Chapter 2.

As it stands, our definition of fib is incomplete because it does not tell us how the sequence starts; to find the third term of the sequence we need to define the first two and the prescription given does not tell us what these are. This is a property of all recursive definitions; we need to start them off by defining **initial conditions**. In the case of fib we traditionally define the initial conditions as

$$\text{fib}(0) = 0$$
$$\text{fib}(1) = 1$$

Hence

$$\begin{aligned}
\text{fib}(2) &= \text{fib}(1) + \text{fib}(0) \\
&= 0 + 1 \\
&= 1
\end{aligned}$$

$$\begin{aligned}
\text{fib}(3) &= \text{fib}(2) + \text{fib}(1) \\
&= 1 + 1 \\
&= 2
\end{aligned}$$

Notice that the two initial conditions given are not unique: there is nothing to stop us using any pair of numbers for the first two terms – it all depends on the process being modelled.

The ability to define a function recursively is a most useful device that permits the action of a function to be defined within a **loop** section of a computer program. For example, we have seen that the $(n + 1)$th and $(n + 2)$th terms of an arithmetic sequence are respectively given as

$$\text{arith}(n) = a + nd$$

$$\text{arith}(n + 1) = a + (n + 1)d = a + nd + d$$

From this it can be seen that

$$\text{arith}(n + 1) = \text{arith}(n) + d$$

That is, the $(n + 2)$th term is given in terms of the earlier $(n + 1)$th term and the common difference. The purpose of this recursive description of arith is that it permits sequential terms to be produced by considering what has gone before rather than by repeated application of a process. As the prescription stands, however, it is lacking because it does not tell us how the sequence starts. To overcome this we introduce the initial condition:

$$\text{arith}(0) = a, \text{ the first term}$$

As soon as we are given this initial term we can now proceed to generate all the other terms in the sequence successively (recursively), as can be seen in the following segment of code:

```
function arith(n : integer) : integer;
begin
  if n = 0 then
      arith := a
  else arith := arith(n - 1) + d
end;
```

Conclusion: The successive terms of every countable sequence can be generated by the recursive application of the function that defines the sequence. To initiate such generation requires initial conditions that define the values of the initial terms of the sequence.

Worked example 8.23

Find the first five terms of each of the following recursively defined sequences:
(a) $f(n + 2) = f(n + 1) - f(n)$; $f(0) = 0$, $f(1) = 1$
(b) $f(n + 3) = f(n + 2) + f(n + 1) + f(n)$; $f(0) = 1$, $f(1) = 3$, $f(2) = 5$
(c) $f(n + 2) = 3f(n + 1) + 2f(n)$; $f(0) = -1$, $f(1) = 1$
(d) $f(n + 3) = f(n + 2) - 2f(n + 1) + 3f(n)$; $f(0) = -1$, $f(1) = 0$, $f(2) = 1$

Solution:
(a) $f(2) = f(1) - f(0) = 1 - 0 = 1$
      $f(3) = f(2) - f(1) = 1 - 1 = 0$
      $f(4) = f(3) - f(2) = 0 - 1 = -1$

(b) $f(3) = f(2) + f(1) + f(0) = 5 + 3 + 1 = 9$
       $f(4) = f(3) + f(2) + f(1) = 9 + 5 + 3 = 17$
(c) $f(2) = 3f(1) + 2f(0) = 3 + (-2) = 1$
       $f(3) = 3f(2) + 2f(1) = 3 + 2 = 5$
       $f(4) = 3f(3) + 2f(2) = 15 + 2 = 17$
(d) $f(3) = f(2) - 2f(1) + 3f(0) = 1 - 0 + (-3) = -2$
       $f(4) = f(3) - 2f(2) + 3f(1) = (-2) - 2 + 0 = -4$

Worked example 8.24

Give a recursive definition of each of the following sequences:
(a) $\text{arith}(n) = 3 - 4n$
(b) $\text{geom}(n) = 6(-0.1)^n$
(c) $f(n) = 1/n$

Solution:
(a) $\text{arith}(n + 1) = 3 - 4(n + 1) = \text{arith}(n) - 4$

Therefore

$\text{arith}(n + 1) = \text{arith}(n) - 4$ where $\text{arith}(0) = 3$

(b) $\text{geom}(n + 1) = 6(-0.1)^{n+1} = (-0.1)\text{geom}(n)$

Therefore

$\text{geom}(n + 1) = (-0.1)\text{geom}(n)$ where $\text{geom}(0) = 6$

(c) $f(n + 1) = 1/(n + 1) = [n/(n + 1)](1/n) = [n/(n + 1)]f(n)$

Therefore

$f(n + 1) = [n/(n + 1)]f(n)$ where $f(1) = 1$

Exercises

8.23 Find the next three terms of each of the following recursively defined sequences:
(a) $f(n + 2) = f(n + 1)/f(n)$; $f(0) = 1$, $f(1) = 2$
(b) $f(n + 3) = f(n + 2)[f(n + 1) + f(n)]$; $f(0) = 1$, $f(1) = 3$, $f(2) = 5$
(c) $f(n + 2) = f(n + 1) - 2f(n)$; $f(0) = -1$, $f(1) = 1$
(d) $f(n + 3) = [f(n + 2)f(n + 1)]/f(n)$; $f(0) = 1$, $f(1) = 2$, $f(2) = 3$

8.24 Give a recursive definition of each of the following sequences:
(a) $\text{arith}(n) = 7n + 2$
(b) $\text{geom}(n) = -9(1/3)^n$
(c) $f(n) = (-1^n)/n$

Functions of two or more variables

Functions of two variables
So far we have considered functions whose domain elements are referenced by a single variable. Functions whose domain elements are referenced by two variables can be defined in a similar manner (Figure 8.15).

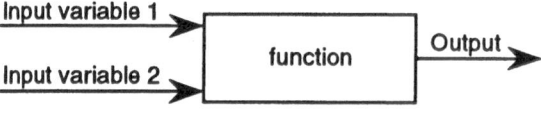

Figure 8.15

For example, in Chapter 2 we considered the problem of defining addition and we found that this could be catered for by defining the recursive process ADD where

$$n \text{ ADD } m = m \qquad\qquad\qquad\quad \text{if } n = 0$$
$$= \text{SUC } ([\text{PRE } n] \text{ ADD } m) \quad \text{otherwise}$$

This prescription can be described by using a function called sum that takes as its input an ordered pair of natural numbers:

$$\text{sum: } (n, m) \mapsto \text{sum}[(n, m)] = m \qquad\qquad\qquad \text{if } n = 0$$
$$= \text{SUC}[\text{sum}(\text{PRE}n, m)] \quad \text{otherwise}$$

where

$$n, m, \text{sum}[(n, m)] \in \mathcal{N}$$

We describe sum as being a function of two variables and its graph is the set of ordered triples G where

$$G = \{(n, m, n + m) \text{ where } n, m \in \mathcal{N}\}$$

Notice that it is not possible to construct a digraph for such a function. Indeed, the only graph that can be drawn in this case is that of a three-dimensional Cartesian graph.

Conclusion: A function of more than one variable can be defined in the same manner as a function of a single variable where the variable is structured in terms of two or more simpler variables.

Worked example 8.25

In Chapter 2 we considered the problem of defining multiplication and we found that this could be catered for by defining the recursive process MUL where

$$x \text{ MUL } y = 0 \qquad\qquad \text{if } y = 0$$
$$= x \text{ ADD } [x \text{ MUL (PRE } y)] \quad \text{otherwise}$$

Describe this prescription by using a recursive function of two variables called prod that takes as its input an ordered pair of natural numbers.

Solution:

$$\text{prod: } (n, m) \mapsto \text{prod}[(n, m)] = m \qquad\qquad\qquad \text{if } n = 0$$
$$= n \text{ sum } [\text{prod } (n, \text{PRE } m)] \quad \text{otherwise}$$

where

$$n, m, \text{prod}[(n, m)] \in \mathcal{N}$$

Exercise

8.25 In Chapter 2 we considered the problem of defining subtraction and we found that this could be catered for by defining the recursive process SUB where

$$n \text{ SUB } m = 0 \qquad\qquad\qquad \text{if } n = m$$
$$= \text{SUC } ([\text{PRE } n] \text{ SUB } m) \quad \text{otherwise}$$

Describe this prescription by using a recursive function of two variables called diff that takes as its input an ordered pair of natural numbers.

Currying

The function sum is a function that maps ordered pairs of natural numbers to natural numbers:

$$\mathcal{N} \times \mathcal{N} \mapsto \mathcal{N}$$

For such a function to be of use within a computer program requires the structured data type of pairs of natural numbers to be defined in addition to the data type of the naturals. The need to construct such structured inputs can severely restrict the applicability of computer software. For example, consider a database that consists of records containing the details of persons living within a certain geographical area. Such a database can be queried, for instance, to find the number of records in the database whose income, age and number of dependants fields have certain predefined values by using the function query, where

query: (income, age, dependants) $\mapsto$ query(income, age, dependants) = number

where

income $\in \{n: n$ is number to two decimal places$\}$ dependants, age $\in \mathcal{N}$

and

number $\in \mathcal{N}$

To use such a function would require the use of the data type of ordered triples of the form

(income, age, dependants)

All databases should be designed to permit the maximum flexibility when querying them, which, according to the above discussion, entails defining all possible data types to encompass all possible queries. If a database has three fields there are six different ways in which ordered multiples of different fields can be defined. This is obviously a very inefficient way to query a database.

A more efficient approach uses a method known as **currying**, in which a function of many variables is evaluated one variable at a time. For example, consider the action of adding 3 to 4 using a hand calculator. We shall define the calculator to be function1 capable of having a number input:

```
function1 : Evidenced by the display of 0
INPUT
      Enter the number 3
      Press the + key to complete the entry
OUTPUT function2 : Evidenced by the display of the number 3 and
                   being ready to add the next input to 3
```

Pressing the + key completes the input but restricts the action of the calculator to one of being only able to add the next number input – this output is another function. The calculator is now defined to be function2:

```
function2 :   Evidenced by the display of the number 3 and being
                ready to add the next input to 3
INPUT
                Press the key 4
                Press the = key to complete the entry
OUTPUT 7
```

Pressing the = key completes the input and produces the output 7. Here we see that

$$\text{function1}: 3 \mapsto \text{function1}(3) = \text{function2} \quad (\text{the output is function2})$$

where

$$\text{function2}: x \mapsto \text{function2}(x) = 3 + x$$

As a second example consider the function

$$\text{sum}: (n, m) \mapsto \text{sum}[(n, m)] = m \qquad\qquad\qquad \text{if } n = 0$$
$$= \text{SUC}[\text{sum}(\text{PRE}n, m)] \quad \text{otherwise}$$

where

$$n, m, \text{sum}[(n, m)] \in \mathcal{N}$$

This function requires ordered pairs to be input. Now consider the modified function sum′ where

$$\text{sum}': n \mapsto \text{sum}'n$$

where

$$n \in \mathcal{N}$$

Here the action of sum′ is to accept as input n and produce as output the function of a single variable sum′n:

$$\text{sum}'n: m \mapsto \text{sum}'n(m) = m \qquad\qquad\qquad \text{if } n = 0$$
$$= \text{SUC}[\text{sum}'(\text{PRE}n, m)] \quad \text{otherwise}$$

where

$$m, \text{sum}'n(m) \in \mathcal{N}$$

The function sum′n accepts m as input and then outputs the result $n + m$.

This process can be extended to functions of more than two variables. Currying a function of n variables entails converting the function of n variables to a function of $n - 1$ variables by evaluating the original function for just one of the n variables. This process is repeated until one ends up with a function of a single variable which can be evaluated in the normal way.

Conclusion: To evaluate a function of more than one variable requires structured data types to be defined. To avoid this complication the function can be curried by evaluating it one variable at a time.

Worked example 8.26

The function max returns the maximum value of two input numbers:

$$\text{max}: (x, y) \mapsto \text{max}(x, y) \quad = x \qquad \text{if } x \geq y$$
$$= y \qquad \text{otherwise}$$

where

$$x, y \in \mathcal{R}$$

Rewrite the process of obtaining the maximum of two numbers as a curried function.

Solution:
This function requires an ordered pair to be input. Now consider the modified function max′ where

$$\text{max}': x \mapsto \text{max}'x$$

where

$$x \in \mathcal{R}$$

Here the action of max′ is to accept as input x and produce as output the function of a single variable max′x where

$$\text{max}'x: y \mapsto \text{max}'x(y) \quad = x \qquad \text{if } x \geq y$$
$$= y \qquad \text{otherwise}$$

where

$$x, y \in \mathcal{R}$$

Worked example 8.27

The function difflen returns the magnitude of the difference in length of two input strings:

$$\text{difflen: } (<\text{string1}>, <\text{string2}>) \mapsto \text{difflen}(<\text{string1}>, <\text{string2}>)$$
$$= |\text{len}(<\text{string1}>) - \text{len}(<\text{string2}>)|$$
$$= n$$

where

$$<\text{string}> \in \{\text{strings}\}$$

and

$$\text{len}(<\text{string}>), n \in \mathcal{N}$$

Rewrite the process of obtaining the difference in length of two strings as a curried function.

Solution:

$$\text{difflen}': <\text{string1}> \mapsto \text{difflen}'<\text{string1}>$$

where

$$<\text{string1}> \in \{\text{strings}\}$$

$$\text{difflen}'<\text{string1}>: <\text{string2}> \mapsto \text{difflen}'<\text{string1}>(<\text{string2}>)$$
$$= |\text{len}(<\text{string1}>) - \text{len}(<\text{string2}>)|$$
$$= n$$

where

$$<\text{string2}> \in \{\text{strings}\}$$

and

$$\text{len}(<\text{string}>), n \in \mathcal{N}$$

Worked example 8.28

The function dateformat returns the display of a date in the format DD/MM/YY from the input of the triple of natural numbers (DD, MM, YY):

dateformat: (DD, MM, YY) $\mapsto$ dateformat(DD, MM, YY)
= DD/MM/YY

where

DD, MM, YY $\in \mathcal{N}$

Rewrite the process of obtaining the difference in length of two strings as a curried function.

Solution:

dateformat′: DD $\mapsto$ dateformat′DD

where

DD $\in \mathcal{N}$

dateformat′DD′: MM $\mapsto$ dateformat′DD′MM

where

MM $\in \mathcal{N}$

dateformat′DD′MM′: YY $\mapsto$ dateformat′DD′MM′(YY) = DD/MM/YY

where

DD $\in \mathcal{N}$

Exercises

8.26 The function min returns the minimum value of two input numbers:

min: $(x, y) \mapsto$ min$(x, y) = x$ if $x \leq y$
$= y$ otherwise

where

$x, y \in \mathcal{R}$

Rewrite the process of obtaining the minimum of two numbers as a curried function.

8.27 The function lensquare returns the sum of the squares of the lengths of two input strings:

lensquare: (<string1>, <string2>) $\mapsto$ lensquare(<string1>, <string2>)
= [len(<string1>)]^2 + [len(<string2>)]^2

where

<string> $\in$ {strings}

and

len(<string>) $\in$ $\mathcal{N}$

Rewrite the process of obtaining the sum of the squares of the lengths of two strings as a curried function.

8.28 The function integersum returns the sum of a sequence of three input integers:

integersum: $(p, q, r) \mapsto$ integersum$(p, q, r) = p + q + r$

where

$p, q, r \in \mathcal{N}$

Rewrite the process of obtaining the sum of three integers as a curried function.

Solutions to exercises

Solutions are given to selected exercises throughout the text.

1.3
(a) Tom is a councillor AND Dick is a councillor.
(b) Petrol is expensive AND diesel is less expensive.
(c) The credit given exceeded the limit AND the quantity delivered exceeded the limit.

1.6
(a) He leaves OR I leave.
(b) That was good news to some people OR that was bad news to some people.
(c) The answer is yes OR the answer is no.

1.9
(a) 5 red is NOT sufficient AND 10 blue is NOT sufficient.
(b) They are NOT wide AND they are NOT narrow.
(c) It was NOT not the best of times AND it was NOT not the worst of times.

1.11
(a) It is August OR today is Thursday.
(b) It is August AND today is NOT Thursday.
(c) It is NOT August OR today is NOT Thursday.
(d) It is NOT August or today is Thursday.
(e) Neither is it August nor is today Thursday.

2.2
(a) If the book is on my chair then I am reading it.
(b) If the sun is shining then I do not carry an umbrella.
(c) If the lecturer is not on time then there will not be enough time to cover the material.
(d) If it is neither time nor appropriate then it is either not time or it is not appropriate.

2.5
(a) Despatch.
(b) Do not despatch.
(c) Despatch.
(d) Despatch.

2.7

The program will print the numbers 0, 1, 2, 3, during which time the proposition Count <> 4·is TRUE. After it has output the number 3 the Count value is increased to 4 and the proposition Count <> 4 is tested and found to be FALSE. As a result execution leaves the loop and the program ends.

2.12

The candidate will fail only if: candidate obtains less than 50% in the examination and is not exempt from the assessment.

3.2
(ia) $\varnothing$, {10}, {100}, {1000}, {10, 100}, {10, 1000}, {100, 1000}
(iia) Set elements as (ia) plus A. Cardinality 8.
(ib) $\varnothing$, {hearts}, {spades}, {clubs}, {diamonds}, {hearts, spades}, {hearts, clubs}, {hearts, diamonds}, {spades, clubs}, {spades, diamonds}, {clubs, diamonds}, {hearts, spades, clubs}, {hearts, spades, diamonds}, {hearts, clubs, diamonds}, {spades, clubs, diamonds}
(iib) Set elements as (ib) plus A. Cardinality 16.
(iic) Cardinality 32.

3.4

The power set of $\{\varnothing\}$ is $\{\varnothing, \{\varnothing\}\}$.

3.10
(a) {2, 4, 6, 8}
(b) {b, d, g, h}
(c) {February, April, May, June, August, October, November}

3.17
(a) $(A \cap B')'$
(b) $A \cup (B \cap C')$
(c) $A' \cap B'$

3.20
(a) (Names beginning with R) $\vee$ ((Worst score < 105) $\wedge$ (Best score $\geq$ 85))
(b) (Best score $\geq$ 85)
(c) (Names beginning with R) $\wedge$ (Latest score < 92)

3.22

2 ADD 3 = SUC ((PRE 2) ADD 3)

         = SUC (1 ADD 3)

         = SUC (SUC ((PRE 1) ADD 3))

         = SUC (SUC (0 ADD 3))

         = SUC (SUC 3)

         = SUC 4

         = 5

3.25

6 DIV 3 = ((6 SUB 3) DIV 3) ADD 1

         = (3 DIV 3) ADD 1

         = 1 ADD 1

         = 2

3.27

(a) After 50 recursions the change in the value of ϕ obtained does not affect the first 11 significant figures.

(b) The number found is a rational number and ϕ is an irrational number.

(c) No – a computer cannot evaluate irrational numbers.

4.2

(a) s

(b) <<e>, <t>>

(c) <s, e, t>

(d) <o, f>

4.5

(a) $\{-1, 0\}$ and $\{p, q, r\}$

(b) {Bradford, Vancouver, Sydney} and {England, Australia, Canada}

4.8

The first elements from A = {Ford, Vauxhall}

The second elements from B = {three-door, four-door}

The third elements from C = {blue, red, white}

(a) $A \times B \times C$

(b) $(A \times B) \times C$

4.11

(a) $4 + 4n$, $2n(n + 1)$

(b) 7^n, $(7^n - 1)/6$

(c) $(-1)^n$, 0 or 1 depending upon whether n is even or odd respectively

4.14

(a) $n + 1$

(b) $1/[n(n + 1)]$

4.16

$(n - 1)! / (n - 5)! = (n - 1)(n - 2)(n - 3)(n - 4)$

4.18

$^7C_4 = 7!/(4!3!) = 35$

4.21

8

5.2

(a) Proposition.
(b) Open sentence.
(c) Open sentence.
(d) Open sentence.

5.6

(a) $\forall x \in \{dogs\}$, x has his day
(b) $\forall x \in \{flowers\}$, x has at least three petals
(c) $\forall x \in \{oranges\}$, x is not ripe
(d) $\forall x \in \{English\ words\}$, x possesses at least one vowel

5.8

(a) $\exists x \in \{buses\}$, x is always late
(b) $\exists x \in \{days\}$, on x your boat will come in
(c) $\exists x \in \{students\}$, x scored full marks
(d) $\exists x \in \{books\}$, a copy of x is in your library

5.10

(a) One of the computers is not a PC.
(b) Someone does not like surprises.
(c) There is a house that is not built out of brick.
(d) One of the voters agreed with the candidate.

5.12

(a) There are no gold mines in those hills.
(b) My luck will never change.
(c) There are no holes in the road.
(d) I shall never go there.

5.14

(a) There is at least one person who has not read all the Waverley novels.
(b) Someone knows all the facts.
(c) There is a map on which all the villages can be found.
(d) There is at least one library book with a diagram on every page.

5.16

(a) There is a dog that does not have his day.

(b) There is a person who does not have a special day.

(c) There is a person who should never come to the aid of the party.

(d) There is a person who is never fooled.

6.2

(a) $((p \rightarrow q) \wedge q) \rightarrow p$ is not a tautology.

(b) $(((p \wedge q) \rightarrow r) \wedge \neg q) \rightarrow (p \rightarrow r)$ is not a tautology.

7.2

(a) $y = x^3$, where $x \in \{-3, -1, 0, 1, 3\}$

(b) (x, y) where x and y are three-digit binary numbers which, in the decimal representation, are odd and where $x > y$

(c) (x, y) where $x, y \in \{A, B, C, D\}$ and where x is before y in the alphabet

(d) (x, y) where $x, y \in \{\text{<four>}, \text{<seven>}, \text{<ten>}\}$ and len $x <$ len y

7.4

(a) $\{(x, x), (x, y), (y, x), (y, z)\}$

(b) $\{(A, A), (B, A), (B, D), (D, B), (D, C), (C, C)\}$

(c) $\{(0, 0), (1, 1), (2, 2), (0, 1), (1, 0), (0, 2), (2, 0), (1, 2), (2, 1)\}$

(d) $\{(a, d), (d, b), (b, a), (b, e), (e, c), (c, b)\}$

7.6

(a) (x, y) where $x, y \in \{\text{bat, cat, fat, hat}\}$ and where the first letter of y is the same as or later in the alphabet than the first letter of x

(b) (x, y) where $x, y \in \{2, 4, 8, 12\}$ and where $x \leq y$

(c) (x, y) where x, y are subsets of $\{a, b\}$ and where x is a subset of y

(d) (x, y) where x, y are binary numbers ≤ 11 and where $x \leq y$

7.8

All of them except (b).

7.10

All of them.

7.12

(a) and (b).

7.15

(a) $(2 + (3 \times 4)) - (10 \div (8 - 3)) = 12$

(b) $((1 + 2) \times 3) - ((6 - (3 + 4)) \times (8 \div (10 - 2))) = 10$

(c) $((3 \times (1 - 2)) + (8 \div (4 - 2))) \div 2 = 1/2$

8.2

(a) journey: destination $\mapsto$ journey(destination) = number of miles to destination

(b) cuberoot: $x \mapsto$ cuberoot(x) = cube root of x

(c) select: brochure $\mapsto$ select(brochure) = holiday

(d) spellcheck: document $\mapsto$ spellcheck(document) = correctly spelt document

8.4

(a) Domain, set of destinations; range, set of mileages.

(b) Domain, set of real numbers; range, set of real numbers.

(c) Domain, set of brochures; range, set of holidays.

(d) Domain, set of documents; range, set of correctly spelt documents.

8.6

(a) journey: destination $\mapsto$ journey(destination) = n, where destination $\in$ {town names} and $n \in \mathcal{N}$

(b) cuberoot: $x \mapsto$ cuberoot(x) = y ,where $x, y \in$ {real numbers}

(c) select: brochure $\mapsto$ select(brochure) = holiday, where brochure $\in$ {available brochures} and holiday $\in$ {available holidays}

(d) spellcheck: document $\mapsto$ spellcheck(document) = c, where document, $c \in$ {d: d is a document}

8.9

(a) Many to one: many lists of numbers can have the same total.

(b) Many to one.

8.11

(a) comp5 = mult2 o sub9 o pow5 where pow5: $x \mapsto$ pow5(x) = x^5

(b) sub9 o cube o mult2

(c) cube o sub9 o mult2

(d) sub9 o cube o mult2 o sub9 o mult2

8.13

(a) hexbin: $b \mapsto$ hexbin(b),where hexbin(b) $\in$ {h: h is a hexadecimal number}

(b) It is not possible to define the inverse function as the original function is many to one. If the departments were such that their numbers were all different then an inverse could be defined as: department: number $\mapsto$ department(number) = <department>

(c) As in the previous question to form an inverse function it must be stipulated that the original function be one to one: recordnumber: amount $\mapsto$ recordnumber(amount) = record number

(d) mult8: $x \mapsto$ mult8(x) = $8x$

8.16
(a) {(a, c), (a, a), (a, b), (b, b), (b, a), (d, b), (d, c)}
(b) $x = y + 2$
(c) [(c, a), (a, a), (b, a), (b, b), (a, b), (b, d), (c, d)}

8.18
(a) $-13, -17, -21$ subtracting 4 each time.
(b) 2592, 15 552, 93 312 multiplying by 6 each time.
(c) $-5/243, 6/729, -7/2187$ multiplying by $-1/3$ and increasing the denominator by 1 each time.
(d) 5.000005, 6.0000006, 7.00000007.

8.22
8, 12, 18,... first term 8, common ratio 3/2.

8.24
(a) $\text{arith}(n + 1) = \text{arith}(n) + 7$: $\text{arith}(0) = 2$
(b) $\text{geom}(n + 1) = (1/3)\text{geom(n)}$: $\text{geom}(0) = -9$
(c) $f(n + 1) = -f(n)[n/(n + 1)]$: $f(1) = -1$

8.27

  lensquare$'$: <string1> $\mapsto$ lensquare$'$<string1>

where

  <string1> $\in$ {strings}

  lensquare$'$<string1>: <string2> $\mapsto$ lensquare$'$<string1>(<string2>)
        $= [\text{len}(\text{<string1>})]^2 + [\text{len}(\text{<string2>})]^2$
        $= n$

where

  <string1> $\in$ {strings}

and

  len(<string>), $n \in \mathcal{N}$

Bibliography

The books cited below are divided into three categories.

Books of general interest whose contents move tangentially away from the subject matter of this book but where an appreciation of this book's contents is of value;

books whose contents are broadly similar and are included to present an augmented and alternative viewpoint;

books that take the subject matter of this and other books to a greater depth and place it firmly within the context of software construction.

General interest

Bartly III, W.W. (1977) *Lewis Carroll's Symbolic Logic*, Harvester Press, Hassocks.

Hofstadter, D.R. (1980) *Godel, Escher, Bach, an Eternal Golden Braid*, Penguin, Harmondsworth.

van Rooten, L. d'A. (1967) *Mots D'Heures: Gausses, Rhames*, Penguin, New York.

Similar

Epp, S.S. (1991) *Discrete Mathematics with Applications*, Wadsworth, Belmont, CA.

Gersting, J. (1993) *Mathematical Structures for Computer Science*, 3rd edn, W.H. Freeman, New York.

Kolman, B. and Busby, R.C. (1987) *Discrete Mathematical Structures for Computer Science*, 2nd edn, Prentice-Hall, Englewood Cliffs, NJ.

Norcliffe, A. and Slater, G. (1991) *Mathematics of Software Construction*, Ellis Horwood, New York.

Woodcock, J. and Loomes, M. (1988) *Software Engineering Mathematics*, Pitman, London.

Deeper

Fenton, N. and Hill, G. (1993) *Systems Construction and Analysis*, McGraw-Hill, London.

IEE (1987) *Essential Mathematics for Software Engineers*, Peter Peregrinus, London.

Ince, D.C. (1988) *An Introduction to Discrete Mathematics and Formal System Specification*, Clarendon Press, London.

Manna, Z. and Waldinger, R. *The Logical Basis for Computer Programming*, Vol. 1, Addison-Wesley, Reading, MA.

Open University (1988) *An Introduction to Formal Methods of Software Development*, Vol. 1, Open University Press, Milton Keynes.

Index